25 FEB 2011

I.L.L.
Date Due:
27 FEB 2017
01285 8516

WITHDRAWN

Please return on or before the latest date above.
You can renew online at www.kent.gov.uk/libs
or by phone 08458 247 200

CUSTOMER SERVICE EXCELLENCE

Libraries & Archives

C333011249

The Star-Crossed Stone

The Star-Crossed stone

The Secret Life, Myths, and History of a Fascinating Fossil

Kenneth J. McNamara

The University of Chicago Press   Chicago and London

KENNETH J. MCNAMARA is a senior lecturer in the Department of Earth Sciences and a fellow of Downing College at the University of Cambridge. His books include *Shapes of Time* and, with John A. Long, *The Evolution Revolution: Design without Intelligence*.

The University of Chicago Press, Chicago 60637
The University of Chicago Press, Ltd., London
© 2011 by The University of Chicago
All rights reserved. Published 2011
Printed in the United States of America

20 19 18 17 16 15 14 13 12 11   1 2 3 4 5

ISBN-13: 978-0-226-51469-7 (cloth)
ISBN-10: 0-226-51469-2 (cloth)

Library of Congress Cataloging-in-Publication Data

McNamara, Ken.
    The star-crossed stone : the secret life, myths, and history of a fascinating fossil / Kenneth J. McNamara.
      p. cm.
    Includes bibliographical references and index.
    ISBN-13: 978-0-226-51469-7 (hardcover : alk. paper)
    ISBN-10: 0-226-51469-2 (hardcover : alk. paper) 1. Sea urchins, Fossil—Collectors and collecting—History. 2. Fossils—Collectors and collecting—History. 3. Collectors and collecting. 4. Paleontology. I. Title.
QE783.E2M36 2010
563'.95—dc22
                                                                                                      2010011100

∞ The paper used in this publication meets the minimum requirements of the American National Standard for Information Sciences—Permanence of Paper for Printed Library Materials, ANSI Z39.48-1992.

For Sue

And these old stones, give them a voice and what a tale they'd tell.

AESCHYLUS, *Agamemnon*

It is the pebble which tells man's story. Upon it is written man's two faces, the artistic and the practical. They are expressed upon one stone over which a hand once closed, no less firm because the mind behind it was submerged in light and shadow and deep wonder.

LOREN EISELEY, *The Star Thrower*

CONTENTS

Prologue *1*

Introduction *5*
1. Awakenings *17*
2. First Collectors *29*
3. Urchins *43*
4. Skeletons *63*
5. Fossilized Memories *79*
6. Maud *93*
7. Augustus Henry Lane Fox Pitt-Rivers *109*
8. Shepherds' Crowns and Fairy Loaves *121*
9. Thunderstones *135*
10. Holy Urchins *155*
11. The Morning Star *169*
12. Snakes' Eggs *185*
13. Star of Destiny *201*
14. Five-Star Attraction *211*
Epilogue *227*

Acknowledgments *233*
Notes *237*
References *251*
Index *263*

PROLOGUE

It was a cold night, the first since summer drifted into autumn. But at least, the young woman thinks, it has been a peaceful one. For the last few days the Thunder God had unleashed his anger on her settlement, which nestled close to the foot of the downs. His black, pendulous clouds had, in their usual belligerent manner, spewed forth hailstones as big as eyeballs and tossed more javelins of fire than she could count. Pulling her cape tight around her shoulders, the woman shivers as she begins her steep climb through the dawn mist.

The woman is not alone. Her child clings to her like a tangle of ivy, pushing his head into the warmth of her shoulder as he tries to stay asleep. He whimpers softly through his mother's cloak while she climbs, steadily and purposefully. Her feet crunch through the frozen grass. The creep of soil down the slope has created green steps on this, the steepest part of the hill. She isn't sure whether this makes her climb any easier or not. Her breathing becomes louder and more labored with each step. Plumes of condensing breath accompany her. Sodden sheep drift in and out of view, scuttering away from her as she disturbs them.

She is not tall, a mere five feet in height. Her dark, braided hair clings to her head, becoming wetter with each moment spent in this freezing mist. As she nears the top of the hill the mist suddenly thins and a feeble Sun makes its first appearance. She skirts a patch of gorse, scattering sunlit drops as she brushes past the wet yellow flowers, and heads straight for a pile of fresh chalk that whitens the top of the hill.

The white mound is stark and cold against the grass that surrounds it, like the pallid belly of a whale floating upturned in an emerald sea. The child begins to struggle and cry, mewing like a kestrel: a shadow that wheels in the brooding sky. The woman unties her cape, throws it down on the grass, and sets the boy on it. Crouching down on her haunches she looks back to where her settlement lies, far below, shrouded in a silent mist.

It has been just a single cycle of the Moon since her last trek here. But here is a place you come to bury your dead, to honor them, or to join them on their journey into the next world. That last time she had followed a long, winding procession as it carried four bodies up the hill on their final journey. One had been the boy's father. It had only taken them a few days to die. Theirs had been desperate deaths; they seemed to fade as blood-wracked coughs tore them apart, until their spirits were released from their gaunt bodies. Bodies that had lived and loved now lie curled and withered beneath this chalk shroud.

How soon, she wonders, before the living join the dead? In the night her child began to cough, and fear shot through her with each little hack. But the day brought release. Perhaps it is nothing. She looks at him, his little hands tunneling through the grass, searching for pieces of gray and white flint. The woman smiles. Already the boy has his father's knack of searching for the best ones. Her man had been a master at crafting the finest, sharpest flint points. No one in living memory had fashioned them as well as he.

He had also been a true son of the Thunder God. Not only was he a great fighter, but he had surely been blessed. It was only he who in the fiercest of thunderstorms would brave the fury and dodge the spears of fire as he scoured the ground for the God's thunderstones. No one, it is said, had amassed such a hoard of these smooth white stones that were seared with a star—the Thunder God's mark. Her man would survive his journey into the next world. With so many thunderstones to accompany him on his long journey to the next life, he would never lose his way. But he had refused to have them buried with him. She must take them, he said. All of them, so that in the future she and the boy would be sure to join him.

She is brought back to the present by the sound of a cry coming from her child. The woman leaps to her feet and rushes across to the lone blackthorn tree to which the boy wandered. But it is not a cry of fear or hunger: it is a cry of triumph. When she reaches him she stops, transfixed by the sight before her. The boy is standing upright, holding his arms aloft, a look of wonder on his mud-smeared face. Grasped in both his tiny hands, raised high above him as though he is about to place a crown on his own head, is the most magnificent of thunderstones, plucked from exposed soil at the base of the tree. Oh yes, he is his father's boy.

She scoops him up in her arms, kissing him so fast and so furiously that soon they are both helpless with laughter. This will be the first thunderstone of many, she thinks, as they bound down the hill, back to her village. Happier than she has been for a long time, her mind is no longer on the small soul that clings to her with all his might as she runs down the steep slope. She is so caught up in the thoughts of all her child will achieve in the years to come that she doesn't hear the little cough that escapes from the boy as he huddles deep beneath his mother's cloak.

INTRODUCTION

On a March day in 1887 the skeletons of a young woman and a child were found on top of a windswept hill in southern England. But this was not a case for investigation by the local constabulary, for the simple reason that the two bodies had lain in their shallow grave for about four thousand years. Little would be remembered today about this discovery were it not for one very strange feature of the burial. Nestling close to the very fragile bones were hundreds of fossil sea urchins—balls of flint engraved with a five-pointed star. All appeared to have been carefully buried with the bodies in their chalky grave at the time of their interment.

Since that day other graves excavated by generations of archaeologists have been found to contain fossil sea urchins. Such discoveries—along with the recovery of these fossils from many other types of archaeological excavations throughout much of Europe, the Near East, and northern Africa—have revealed that people have been collecting fossils for an extraordinarily long time. Often these fossils show signs of having been altered in some way by their ancient collectors, in some cases many thousands of years ago. It may be just a single fossil that is discovered, but every so often a paleontological treasure trove is found in which hundreds, or in one case thousands, of fossil urchins pour forth from dusty graves.

It would seem that many aspects of our behavior haven't really changed much over tens of thousands of years, for some of us still like collecting fossil sea urchins. These days, though, rather than popping them into a grave, we put them in museums—safe havens for objects accumulated by those with the lust for collecting. We may like to comfort ourselves with the thought that we are far more "civilized" than our jut-jawed, hairy Stone Age ancestors. But is our habit of carefully putting objects like these fossils into serried ranks in cabinets that sleep deeply inside the bowels of our museums really very different from what our ancient ancestors were doing? Their museums were the graves in which they buried their dead, or the houses in which they lived, or the shrines and temples at which they worshipped. For here, too, fossil urchins sometimes took pride of place.

In this book I will use this rather strange object, a fossil sea urchin, to investigate how our collecting habits, and other aspects of our behavior, haven't changed much from those of early humans who lived not just thousands, or even tens of thousands, but hundreds of thousands of years ago. For as we shall see, they, too, seem to have had an inordinate fondness for fossil sea urchins, and been bitten by the collecting bug.

For the last three hundred years or so, most people have accepted the view that such fossils represent the petrified remains of animals that once lived millions of years ago. But what did these prehistoric collectors make of them? Sports of the devil? Gifts from the gods? Why did they bother to collect them? And more important, what drove them to so often bury them with their dead? Not something we do these days. Yet long before they began tucking up their dear, departed friends and relatives for eternity with a cache of fossils, people had been making use of them in other, more practical, ways in their daily lives. But these were not the kind of people that you and I would recognize, for it is not just our species that has had a protracted propensity for collecting these fossils. Even other species of our genus *Homo*, living hundreds of thousands of years ago, were fascinated by them. But just what was it about these particular fossils, which millions of years ago had been sea urchins plowing through the muds of ancient seas, that made them so attractive to so many people, so long ago? And what, if anything, can this tell us about the evolution of the human mind?

Archaeology is more than just digging up fragile bones and mud-encrusted fragments of pots. It is the methodology for studying traces of past human behavior from the remains of objects that have been constructed or modified.[1] While studies of trace fossils, such as footprints, can provide compelling insights into the physical behavior of our distant ancestors, studying archaeological traces has the potential to unravel the thought processes of people who lived long ago. Some archaeological objects may shed light on the more mundane aspects of people's lives, such as what they ate, where they lived, how they clothed themselves, and so on. But there are a small number of objects that, either from their archaeological context or the way in which someone modified them, can illuminate how people might have behaved thousands of years ago. In other words we can gain some small insights into how they might have thought for a brief moment in their lives. We might, then, even begin to understand a little about how individuals tried to come to terms with the complex world into which they had been born.

So how can we ever hope to know what someone who lived hundreds of thousands of years ago thought? One way that we might be able to delve into the thought processes of people who lived long before written records would have fossilized their thoughts for future generations is by looking for clues they left behind in the way that they manipulated the environment they inhabited. We see their reverence for the dead in their burial practices. We can look through their eyes at the wildlife they shared their lives with through their paintings and engravings. How they built their dwellings tells us a lot about how they would have interacted with one another. Yet can we ever hope to unravel what an individual was really thinking? This is a hard enough challenge when we are looking at members of our own species that lived up to two hundred thousand years ago. But what of the hominid with whom our species coexisted, *Homo neanderthalensis*, or what of even our ancestral species, *Homo heidelbergensis*? Is there any way that we can unlock their minds and see through their eyes? I think there is.

From the time the first species of the genus *Homo* evolved more than 2 million years ago, they and all subsequent species increasingly manipulated their environment. We know this because they began to use objects extracted from the environment. Typically only the most resilient items, like stone tools, have survived the destruction of time. But what these objects reveal is that the power to reason must have developed very early in our evolutionary history. The earliest species of *Homo*, *H. habilis*, had probably learned that a broken stone could be used to cut, to chop, to scrape, even to kill. As the brains of their descendants became larger and increased in complexity, their tools became more sophisticated. Then, about 1.4 million years ago the mind made what is arguably its most important evolutionary breakthrough: it developed the ability to understand and appreciate the concept of symmetry. The mind had begun to think in the abstract and to conceptualize. And with this ability might have come an inquisitiveness about the world in which these people had been born, and in which they lived and died. Where did they come from, why were they here, and after death, where did they go? And how to explain this amazing world around them: Who created the thunder and lightning that so terrified them? What could make the land on which they walked periodically shake so violently, or have the power to spew burning gobbets of rock high into the air?

We know that an understanding of symmetry and the acquisition of abstract thought appeared quite early in our evolutionary history because of the beautifully crafted stone implements, known as Acheulian tools, which display perfect bilateral symmetry. Such tools would have been the basic implement used by *Homo erectus* and *Homo heidelbergensis* for cutting, scraping, and chopping for at least one million years. Although most were made to be used for such practical purposes, some are so small, and others so large, that it is hard to imagine that they could ever have been fashioned to be used as practical objects. The only alternative explanation for the manufacture of such objects is that they must have been constructed for the simple pleasure of crafting these early manifestations of the human creative imperative. What, though, drove this quest for producing an object of simple, yet profound symmetry? Was this just an inevitable consequence of the evolution of our increasingly more complex brain and greater cognitive powers? The world in which these early species of *Homo* lived abounded with symmetrical shapes, from the flowers and fruits that they gathered to eat, to the animals that they hunted, and to the animals that on occasions hunted them. Evolving first the ability to mimic this symmetry, and second the desire to carry it out, must surely have been a pivotal moment in the evolution of our cognitive abilities.

Having evolved the ability to think in the abstract and reflect this in the objects they fashioned from material in the world around them, our ancestors would have begun to question their very existence, and their place in the cosmos. And with the development of this abstract thought would have come the first glimmerings of an interest in, and an attraction to, other types of symmetry and patterns—patterns seen in curious structures in the rocks they gathered to make their tools. From this burgeoning imaginative mind sprung forth all manner of wondrous mythological beings: creators of the Earth, providers of life-giving rain who inhabited the star-spangled firmament. And from this it was just a small step to the more formalized concepts of spirituality, of religion—all interwoven in an evolving nexus as this nascent abstract mind flexed its cerebral muscle along the relentless pathway of time.

Then into this mix rolled a flint ball. This was no simple flint ball but one engraved with a five-pointed star. Today we call it a fossil sea urchin. Once the mind's eye had grasped and embraced the concept of symmetry, it latched on to this pentamerous shape that combined the appeal of both bilateral symmetry and a strange form of radial symmetry. It was almost as though some deity with a quirky sense of humor thought, *OK, guys, see what you make of these*—and flicked these stones like cosmic marbles into the world of humans. Sometimes they were as perfectly round as an eyeball or shaped like a heart. At other times they resembled one of the simple huts in which the people dwelled. But in all cases they were seared with the same mark—the five-pointed star.

In this book I will argue that it is possible to use this one strange and seemingly obscure fossil to unlock the ancient mind and to trace how it has evolved, not only in our species, but even in species ancestral to our own. Our bones may hold the key to the evolutionary history of our bodies, but it is the scattering through time of objects that we have treasured, either created or collected for nearly half a million years, that has the power to reveal the secrets of the evolution of the mind. The evolution of abstract thought, of aesthetics, mythology, spirituality, and concepts of fate and destiny, are all in some small way encapsulated in the little stone balls that we now call fossil urchins.

As a species, we are not alone in the animal kingdom in collecting and hoarding objects. While our quarry may be anything from paintings to books, fossil sea urchins, Victorian teapots, or seventeenth-century Bolivian handkerchiefs, almost without exception all other animal species that carry out this acquisitive habit do so for some more practical reason than that it provides them with a rather nebulous degree of satisfaction. The Australian bowerbird (*Ptilonorhyncus violaceus*), for instance, collects whatever blue objects it can lay its beak on. It does so, however, for a purely ulterior motive: to impress the females. But do we collect to impress other members of our species, or only ourselves? Just why should we gain satisfaction from accumulating objects? And if we have been doing this for countless generations—as witness the thousands of museums around the world containing every type of natural or man-made object imaginable, reflecting our collective urge to hoard—have humans, in fact, been as possessive of objects for as long as our species has existed? Exactly what can the apparent evidence that our species (and maybe even earlier species of humans) collected fossil sea urchins tell us about ourselves?

If we think about all the species of animals and plants that have ever been described and given names, the diagnosis of *Homo sapiens* is arguably the vaguest. When a species new to science is formally described, it carries a diagnosis that sets out those features which are unique to the particular species. A species of shark, for instance, may be diagnosed by the size and shape of its teeth and the number of denticles present on the biting surface, along with the shape of the body and so on. For humans, however, Linnaeus (who formally described us in 1758) defined the genus *Homo* as *Nosce te ipsum*, or "Know thyself."[2] Interestingly, William Turton, in the English translation of Linnaeus's work written in 1806,[3] omitted this. Our species, though, *Homo sapiens*, is diagnosed as being "diurnal, varying by education and situation." Such a diagnosis would, I am sure, fail to impress the editor of any modern-day scientific journal dealing with taxonomy (the scientific description and naming of new species).

If someone today were to set out to provide a slightly more embellished diagnosis than the one Linnaeus gave, it is likely to be very different. One characteristic feature that could be included, and that perhaps fits Linnaeus's all-embracing description, would be our propensity to collect, acquire, gather, or otherwise accumulate objects for no apparent practical reason, other than our inquisitive nature. As Linnaeus remarked in the preface to his *Systema Naturae*: "Man, always curious and inquisitive and ever desirous of adding to his useful knowledge; among other sources of amusement and instruction, is naturally led to contemplate and to enquire into the work of nature."[4]

Such an interpretation would suggest a genetic cause for this collecting condition. But there are some who argue that in the "nature/nurture" collecting debate it is all nurture (or more specifically a lack of nurture during development). In his book *Collecting: An Unruly Passion*, Werner Muensterberger argues strongly for the impact of an individual's early upbringing in determining whether or not he or she gets bitten by the collecting bug.[5] In his view, collecting is a compulsive act, molded by irrational impulses that arise from an emotionally unstable upbringing. Thus, the object comes to represent something that will provide emotional support. The collector assigns power and value to the objects, virtually akin, so Muensterberger argues, to a religious fervor. Irrespective of the type of object that is collected, Muensterberger believes that it is the emotional state that collectors achieve by surrounding themselves with magically potent objects that drives them on. But to my mind, it is not just the acquisition of the object; it is the hunt to find it that provides the impetus. It is the hunt that matters, whether for porcelain, fossils, or beer-bottle caps. There is a preoccupation with the challenge—an emotional hunger to collect.

Gathering an object may generate some degree of satisfaction—a warm inner glow that yet one more Roman coin has been added to our burgeoning collection. But for most people it is the hunt that matters most. While our ancient ancestors had plenty on their plate, so to speak, hunting for food in order to satisfy their inner being's nutritional urges, perhaps humans' insatiable appetite for collecting the seemingly most useless of objects satisfies a deeper craving: food for the mind, rather than for the body. And perhaps as our ancient ancestors fulfilled this more cerebral urge they developed a need to collect certain symbols, such as a five-pointed star. Food for the mind could have evolved into food for the soul.

Bones, urchins, and stars—they are a recurring theme in this book. Until the attitude that frowned on the practice of burying goods with the dear departed became prevalent during the Christian era, for thousands of years people had been placing all sorts of items in their loved ones' graves, much to the later delight of archaeologists. In addition to objects of personal adornment, such as clothing and jewelry, the most common articles buried with the dead were those deemed to be of practical use in the afterlife: weapons, animals, pots, bowls, tableware, and such like. Different meanings may be attached to the significance of these funereal collections: they may tell us much about the personal items that the deceased treasured during their life. However, the nature of the grave goods might be indicative of what those who buried the dead thought they would need for their journey into the afterlife.

For the most part it is hard to imagine that fossil sea urchins had any apparent practical purpose. Their presence in graves like those of the four-thousand-year-old woman and child indicates that either the fossils meant a lot to the deceased when they were still alive, or in some way they had a meaning to them in the afterlife. So, to unravel what fossil sea urchins meant to people who lived thousands of years ago—and why they thought it so important to collect them and at times to be buried with them—it is necessary to extract every scrap of evidence from these occasional, often strange, archaeological finds. It is important to look for patterns, such as the occurrence of an axe or dagger with a fossil urchin in a cremation, and also other associations. For example, fossil urchins have been found in archaeological contexts that are directly associated not with the dead, but with the living. The discovery of a cache of fossil urchins in the remains of a settlement that stood in the county of Dorset, England, between the first and fourth centuries is a particularly potent example. These fossils appear to have been deliberately buried under houses a number of times during this period.[6] No written record exists of why people did these things, but we can possibly glean some insight into their motives from oral history, in the form of folklore. It may even be possible to tie this evidence in with certain mythologies, thereby allowing a greater understanding of people's motives for reburying the fossils that had previously been unearthed.

The association of these fossils with the living is shown in other, more tangible ways. There is clear evidence that some urchins were deliberately collected thousands of years ago from specimens found in many parts of Europe, in the Near East, and in Africa: the fossils have been artificially altered. Some have had a hole drilled through them, while others had been scraped or artificially colored. Perhaps the most striking is a fossil urchin found in Egypt on which hieroglyphs of the name of its finder and where he found it were inscribed more than three thousand years ago.

This book tells the story of the fossil sea urchin–collecting habits of early peoples of Britain, much of northern Europe, the Mediterranean region, the Near East, and parts of Africa. It also investigates how these star-crossed stones fascinated different societies for literally hundreds of thousands of years, from the interest shown by early Paleolithic people who shared a northern European landscape with lions and elephants, to those people thousands of years ago who seem to have incorporated them into their spiritual beliefs. Using evidence from archaeology and folklore, it is possible to reconstruct the myths that may have grown up around these fossils—myths that may now be represented merely by a folk name. Just what myths were transmitted from generation to generation about stones known as shepherds' crowns or fairy loaves? And how can we reconstruct these faded myths that exist merely in the shadow of a name?

In this book I also examine the nexus that exists between these fossils and the symbolism of five-pointed stars. These days it is hard to escape this star. It spangles our clothes. It clings to our footballs. It takes pride of place on more than fifty national flags. The marketing moguls have seen to it that it appears in myriad places: on the packaging of our food, on our beer bottles and the logos of our coffee shops. It may even tattoo our bodies. And at Christmas it becomes almost impossible to walk down a street in any city center without being visually assaulted by this heavenly symbol every few seconds as it glitters at us from decorations and merchandise, all shouting, "Buy me! Buy me!" And all the while it sits smugly on top of the Christmas tree, the poor angel having been unceremoniously demoted to one of the lower branches.

As I began delving, some years ago, into the misty world of the fossil-collecting habits of people who lived thousands of years ago, it soon struck me that there are many parallels between why people wanted to collect these star-marked fossil urchins and pop them into their graves, and the interest that, for at least five thousand years, people have shown in drawing the symbol of the five-pointed star itself. For instance, while people were putting the fossil in their graves in many parts of Europe, five-pointed stars were being plastered on the ceilings of burial chambers in ancient Egypt. Some fossil urchins collected in the Near East over ten thousand years ago have been artificially altered in such a way as to enhance the close similarity between the symbol of the five-pointed star and the human form, a relationship personified by Leonardo da Vinci's drawing of the Vitruvian Man and recognized for thousands of years.

The similarity between the fossil and the star is also evident in some European folklore, indicating a long history of using fossil urchins for their perceived apotropaic properties—that is to say, their ability to ward off evil, in exactly the same way that the five-pointed star pattern was often used in medieval Europe. Folklore and mythology also reveal a link between the way both fossil urchins and five-pointed stars were associated with the heavens—for instance, in Norse mythology the fossils were thought to have been hurled to Earth during thunderstorms by Thor, the god of thunder, weather, and crops. These days the five-pointed star is symbolic of the pinpricks of light that pepper the night sky and which we call stars.

But what was it that triggered this mental link between the five-pointed pattern and a bright star in the heavens? To find the answer we must look not to the sky but to the ground. For thousands of years people have been finding and collecting stones to turn them into tools. In so doing they have occasionally encountered these little star-crossed stones. Surprisingly, of all natural objects it is these rather strange fossils more than any other that seem to have fascinated people for the longest period of time. By searching back far enough through the archaeological record of fossil sea urchin collectors, from Paleolithic times hundreds of thousands of years ago right through to the present day, it may be possible to unravel this link between the five-rayed star that sits astride the fossil and the stars in the sky.

The archaeological record shows not only that people collected fossil sea urchins for thousands of years, but that they often did so in huge quantities. But why? What could it have been about these fossils that so attracted them, apart from the five-pointed star? They put them in their houses, buried them with their dead, made tools from them, drilled holes through them, painted them, and put them in necklaces. Could these little star-crossed stones, the fossilized remains of creatures that lived millions of years ago, perhaps be the very source of inspiration for the ubiquitous symbol of the five-pointed star? How different is the desire of someone today who wants to wear a tee shirt emblazoned with a five-pointed star, from our ancestor from four hundred thousand years ago who fashioned a tool from a piece of flint emblazoned with exactly the same symbol? And could, perhaps, our shared fascination with the five-pointed star indicate that maybe the minds of our distant ancestors weren't really so very different from our own? Have all aspects of the mind undergone a steady evolution over the last few hundred thousand years, or are there some fundamental features (such as our ability to conceptualize and imagine) that, once evolved, have changed little in all that time? Can the five-pointed star tell us?

To unravel the critical links between fossil urchins and the star symbol, it is necessary to travel down disparate pathways; many answers must be found for many questions. What can Norse mythology tell us about the Vikings' association of fossil urchins with hand axes? What drove people five thousand years ago to construct a burial chamber twenty-two yards (20 m) across for the sole purpose of placing a single fossil urchin in a box at its center? Why were a Roman emperor's and an Egyptian priest's lives both touched, albeit in very different ways, by fossil urchins? Why, ten thousand years ago, did people in the eastern Mediterranean region apparently view these fossils as fertility symbols? And what prompted a medieval church builder in England to frame a window with a collection of fossil urchins?

In revealing the long-term fascination people have held for these fossils, we owe a debt to five men, each of whom plays a significant role in this story: Cecil Curwin, Worthington Smith, the impressively named Augustus Henry Lane Fox Pitt-Rivers, Herbert Toms, and John Henry Pull. Their archaeological endeavors spanned the late nineteenth and early twentieth centuries. The first three, because of their diligence, played an important role in recognizing the significance of the fossil sea urchins that they discovered at their respective archaeological excavations. The fourth, Herbert Toms, was instrumental in recording the fading folklore of these fossils in early twentieth-century England. The last of the five, John Henry Pull, not only found fossil urchins in a number of archaeological excavations, but also, like Toms, explored their recent folklore in an effort to discover why people living in one small area in southern England had been collecting these fossils four thousand years ago.

The lives of three of them—Curwen, Pitt-Rivers, and Toms—were interwoven. Pitt-Rivers employed Toms and inspired him with a love of archaeology; Toms, in turn, fostered and encouraged Curwen's archaeological exploits. The lives of Curwen and Pull were also interwoven, but in a less than auspicious way: their relationship led to Pull's work being almost totally forgotten by modern archaeologists until recently.

It is because of the efforts of people such as these that we are able to hear faint whispers about these little star-crossed stones—whispers that have traveled like half-forgotten memories through vast tracts of time in the form of myths and legends as people changed from hunter-gatherers to cultivators of the land. Listening carefully to these distant societal murmurs, we may just be able to comprehend something about this ancient preoccupation with fossil sea urchins and the myths they engendered. And herein may also lie the root of an ongoing fascination that people have had for thousands of years with the symbol that brazenly brands these rather strange little fossils: the ubiquitous five-pointed star.

> This was the echinoderm, or petrified sea-urchin. They are sometimes called tests (from the Latin testa, a tile or earthern pot) ... Tests vary in shape, though they are always perfectly symmetrical; and they share a pattern of delicately burred striations. Quite apart from their scientific value ... they are very beautiful little objects; and they have the added charm that they are always difficult to find. You may search for days and not come on one; and a morning in which you find two or three is indeed a morning to remember.
>
> JOHN FOWLES, *The French Lieutenant's Woman* (1969)

1

Awakenings

High on the northern part of Salisbury Plain sits the tiny village of Linkenholt—a few houses, a dairy, the old manor house, and a church. It is early October, 2003. The Sun shines on the churchyard, but meekly, its strength whipped away by the icy northerly wind that sweeps across stubbled fields. It's a pretty church. The walls are made from dark gray and white knuckles of flint, sweated from the Earth, and then collected from the nearby fields. The colors of the flint are mirrored by the black and white wood of the tower, above which rises a conical, shingled spire. St. Peter's of Linkenholt is like many of the flint churches in the remote villages that nestle in the dry valleys of this chalk downland. But this one sits proudly, high on a ridge, the flinty fields fading away in all directions. The flint has fed the people of Linkenholt for thousands of years: they have grown their crops in its biting shards, while their sheep have fed on the springy grass that managed, somehow, to grow in it. And it has fed the walls of their houses, their church, and their dairy.

Little moves in the churchyard apart from a few pied wagtails snapping up unsuspecting insects. But 132 years ago this churchyard was a much less soporific place. Then, it was crowded with more than a hundred people.

18   Chapter One

The village had probably never seen so many people together in one place in its entire existence. They were here in 1871 to celebrate the consecration of the church, which had just been erected on this new site, a couple hundred yards from where its predecessor had sat for almost seven hundred years. But the new structure wasn't a complete break with the past, and with the old traditions. A few choice parts of the original medieval church had been retained. It might well have been easier to build it all again from scratch, but there were some things that had been so important to the people of this village for so long that they couldn't bear to give them up. The stone arch around the door was one. After all, their ancestors from time immemorial had passed through it on their way to be christened, married, or buried. The bowl of the font that had baptized every child since the time of King Richard I was another. And there was a third—a little window. But

Reconstructed medieval window on the north side of St. Peter's Church, Linkenholt, Hampshire, decorated by twenty-two flint fossil urchins. Author photo.

this wasn't one that took pride of place on the sunny side of the church where it could be easily seen by the parishioners as they entered every Sunday. No, this was a simple, unprepossessing, round-headed window, a mere hand's span in width, and a little over five spans long, hidden away on the dank north side of the church. Few would ever go and see it. Here is a place where the Sun has never shone: it is a place to be visited neither by the living, nor by the dead, who aren't even buried here. Some, long ago, are said to have called it the devil's side of the church.

Turning the corner of the building, passing from bright sunlight into perpetual shade, I disturb a flurry of pheasants that whirl into the air. So there is some life here after all. But such activity is rare. The walls of the church here are pale green. A thin layer of lichen is the only permanent life-form, imperceptibly growing on the cold, damp flints. The window is so small that in this eternal twilight it lets little light into the church. Why did the rebuilders bother to keep it at all? It's really not very attractive. But maybe it wasn't the window at all. Perhaps it was what surrounded it: yet more flints. But these flints, like twenty-two pale green hens' eggs, are most unusual: etched onto each of their domed surfaces is a five-pointed star. They are fossil sea urchins. Seven hundred years earlier they had been plucked from the flinty fields that surround the village and, for some strange reason, cemented with great care around the little window.

About 70 million years ago these urchins had crawled through the chalky mud of a warm sea that covered much of what is now northern Europe. They died, and their soft tissue was eaten by myriad hungry marine organisms. Into their empty shells oozed a silica-rich jelly, which eventually hardened to form a perfect flint cast of the inside of the urchin. For tens of millions of years they lay cocooned within their chalky grave. But the despoliation by water, ice, and wind relentlessly shaved layers of chalk from the Earth. Plants did their best to arrest this merciless march of erosion by growing on the thin, flinty soil that covered the chalk. But then a species of mammal arrived that had learned how to scrape off the trees, shrubs, herbs, and grasses that coated the hills.

One day, about fifty years ago, one of these humans was plowing the flint-riddled soil of Linkenholt when he noticed something he was more used to seeing on Sunday mornings. *Ah. Shepherd's crown*, he thought. He reined in his horse, bent down, and picked up what looked like a flint egg. Like many who had plowed these fields for hundreds, maybe thousands, of years, he couldn't resist keeping the flint. Popping it into his pocket, he thought no more of it for the rest of the day. The evening was getting cold. Walking up the path to his front door, he put his hands in his pocket

Fossil urchin *Echinocorys scutatus* collected by Mr. Alan Smith and known by him and his wife, Betty, as a "shepherd's crown"; from a field in Linkenholt, Hampshire, and kept by the front door of his cottage. Fossil length 1.9 inches (5 cm). Author photo.

to warm them up. As his hand touched something smooth and cold he remembered the fossil urchin. He took it from his pocket and dropped it in the flowerpot by his front door, like he had done a number of times before. As he closed the door behind him the last gasp of the evening Sun slipped from behind a cloud and lit up the star on the stone egg.

Hundreds of miles to the east of Linkenholt, and some four hundred thousand years earlier, someone else is scouring the chalk downland for flints.

But this person is not quite human. A little smaller and stockier than you or I, he has prominent eye ridges perched on top of a large face that slides down to a powerful, protruding jaw. This is a descendant of *Homo erectus*, a highly successful species that more than a million years earlier had spread out of Africa: first into Europe and western Asia, then across to China and as far east as Java—maybe even beyond. Not quite *Homo sapiens*: some scientists call it *Homo heidelbergensis*.

He keeps an eye to the ground, scavenging for good, hefty flints to knap into tools. Here, high on a downland ridge, is not the best of places for flint, though. The river gravels are better. But he is up here on his way across to the next valley. He is also keeping a wary lookout for some of the less than companionable fellow mammals: the lions, wild boar, and rhinoceroses that roam this area. Down in the valley herds of fallow deer feed on the lush meadows; a family of straight-tusked elephants makes its way to the river. The valley echoes with their mournful trumpeting. An Ice Age had come, and an Ice Age has gone. Others had yet to spread their white shrouds over the land.

Like other early humans for hundreds of thousands of years before him, this ancestor of modern *Homo sapiens* would appear not to have been very imaginative in the way that he crafted his tools. But the fist-sized flint hand axe that he makes, shaped like a huge teardrop, serves his purposes very well. It is crafted to fit neatly into the palm of his hand and is flaked to razor-sharp edges on both sides. Limited in its design, certainly, but this axe was the Swiss army knife of our ancestors for nigh on a million years. It would probably have been used to skin animals, to chop up the carcasses; grub up roots, cut wood, and chop down trees. Why make anything different, when a single tool can do all this?

He slows down as he notices some large trees, uprooted in a recent storm that wreaked so much havoc. Soil and rocks litter the tendriled feet of these fallen behemoths. Worth checking for flints. Most are probably too small for his purpose. But tucked beneath one of the roots is a larger chunk. It is about the size of a fist, though a little longer and thinner, and dirty brown-gray in color. It would make a good tool. He pushes it with his foot and it rolls away from the tree roots.

As he bends down to pick it up, something on its surface catches his eye—a strange shape, round like the Sun and raised above the surface of the rest of the flint. He runs his rough fingers over its smooth, round surface. As he does so he feels what seem like scratches running across it. He spits on it, and rubs the chalk dust off its surface. Looking more closely, he sees them: five lines, spreading out from the middle of the circle like the main struts of a cobweb. But there is no spider sitting in the center of its web here. It is a strange flint, this one.

Although it takes great skill to turn a rock such as this into a hand axe, for him it was unlikely to have been difficult—knapping away at one edge of the big flint, paper-thin slivers, sharper than a scalpel blade, fly away in showers of sparks. Soon he has done all that he can—another teardrop-shaped tool. Well, almost. He has created what we now call an Acheulian hand axe—the trademark tool of *Homo erectus* and *Homo heidelbergensis*. Although limited in technology, this tool is special because of the bilateral symmetry it possesses, reflecting, perhaps, a mind capable of conceptualizing. This one, though, that today sits in a drawer in Liverpool Museum, is different. Only one edge had been sharpened; the other was left untouched. The last blow had flaked a sliver off this circular object with its

Early Paleolithic Acheulian hand axe with fossil sea urchin (*Conulus*), probably made by *Homo heidelbergensis*; from Swanscombe, Kent. Specimen in Liverpool Museum. Axe length 4.3 inches (11 cm). Author photo.

five-rayed star pattern. Why was it left unfinished? Why was the imperative to produce this bilateral shape not realized? Did our ancient ancestor reason that trying to sharpen the other edge would only have resulted in the attractive pattern being destroyed? Anyway, what did it matter if it only had one edge? The axe would have rested comfortably in his hand. Even though a small piece of the strange five-stranded web had chipped off, when he held the flint his thumb would have fit neatly into where the lost piece had been, his other fingers curling around the smooth side of the axe. It was almost as though his hand had grown this cutting edge—a cutting edge marked with this curious pattern.

Eagles die, and a beech tree falls. Leaves shrivel, but fungi flourish. Beetles gnaw the bark, and fungi's strangling tendrils consume the wood. Soon the tree has faded into the soil, little more than a faint memory—a wan shadow on the ground. Then even that has gone. Days get cooler. Days get colder. Flints, the only remnants left by these waning people, pile up in streams and in the bottoms of rivers. The sands and gravels cover the lost tools of *Homo heidelbergensis*, hidden, lost to the world. Ice smothers the land for what seems an eternity—almost as permanent as the flints.

Eagles fly, and beech trees sprout. The ice fades, and the land greens. Taller people appear. Cleverer people. Eventually they discover that flints have uses other than simply as hand tools. Now that there are tools of iron, who needs tools of flint anyway? But flints can be used to build walls and dwellings, schools and places of worship. Men dig large pits in the gravels, and move thousands upon thousands of flints. A different time, a different species of human.

It is the early twentieth century, and another man is looking through the flints in a quarry in Swanscombe in the county of Kent. Since they were first found in the 1880s, the quarry has yielded more Paleolithic flint implements than any other site in Britain, in excess of sixty thousand, along with three pieces of hominid skull. They belonged to the so-called Swanscombe Man who had evolved beyond *Homo erectus*, but had not yet become modern *Homo sapiens*. The man is searching through gravels from the middle part of the quarry. He spots one that he is looking for— an Acheulian axe: hand axes that in their shape mimic, in many ways, a human hand minus the thumb. With the evolution of symbolic thought in early species of *Homo*, the Acheulian hand axe may have been subconsciously manufactured in this shape because it symbolized a third hand— but a hand with edges so sharp they can cut and skin carcasses. Unlike our asymmetrical hand, this artifact has perfect bilateral symmetry. Yet this one is different. He spots it right away, for it carries on one side, engraved like a Paleolithic logo, a five-pointed star set within a perfect circle—a fossil sea urchin. And the flint has been worked only along one edge.

Trying to delve into the mind of someone who lived nearly half a million years ago is clearly no easy task; impossible, perhaps. Did this individual really select this flint simply because of the fossil urchin? If so, was the underlying urge the same as that which drove people to pick up flints around Linkenholt in far more recent times, and to put them in the walls of their church and in a pot by their front door? If so, it has profound implications for our understanding of how the ancient mind worked.

Although the Swanscombe hand axe no doubt attracted the attention of the more recent finder, what was it about the flint that attracted our paleocollector to it? Dorothy Downes, formerly a curator at the Liverpool Museum where the hand axe is now kept, was in little doubt. As she reported to the paleoanthropologist Kenneth Oakley, who first described it: "Although the fossil has been slightly damaged by the removal of a flake, it was evidently intended to be the central feature of the hand-axe."[1] Certainly, it could be argued that the presence of the fossil sea urchin was a matter of pure chance. On the other hand maybe the shape of the fossil and its strange symmetry caught the paleocollector's eye. If this was the case, what was it about the five-pointed, radiating star blazoned across the fossil's surface, like a forlorn starfish splayed across a rock, that attracted him? Could this person have been driven by the self-same fascination that we have with natural objects today—especially ones with a strange pentagonal symmetry? If so, then this chunk of flint attains almost monumental importance, for it would mark one of the earliest pieces of evidence that we have of an object being collected and kept for some reason other than purely utilitarian. It would have been not just a case of "I need," but, more potently, "I want." The fossil would therefore have appealed to a nascent aesthetic sensibility—an awakening, perhaps, of the perception of what might, in the broadest sense, be construed as beauty. Our beetle-browed, hirsute, jut-jawed ancestors might not have been quite so thuggish as we generally consider them to be. Although their table manners might have left a little to be desired, they appear to have appreciated a good fossil when they found one.

We often think of the birth of art as beginning a little over thirty thousand years ago, when people living in what is now southern France began daubing the walls of caves with exquisite drawings of animals, such as horses, aurochs, and deer. These paleoartists had obviously developed what has been called visual symbolism.[2] Four cognitive and physical processes are recognized in this development: making visual symbols; classifying an image as a type of symbol; communicating information about the symbol; and giving the image a meaning. Yet visual symbolism should also include receptivity on the part of an observer: an "appreciation" of the drawn symbol, or even of the pattern observed in nature. I cannot draw to save my life, but I can appreciate a painting by Monet or Turner. So, too, those living long before people first put ochre to rock face could well have developed what can only be termed an attraction to, or appreciation of, a shape, a form, or a symmetry that they encountered in their everyday world.

Our appreciation of a painting or a sculpture may lie as much in its form as in its color. In its simplest manifestation, this is the orderliness of circles, portraying a symmetry and a completeness—a satisfying closure. Or the form may be a spiral, or radiating rays. It is an appreciation of these qualities in the natural world as much as in the created world, where handprints or an exquisite head of a horse painted on the wall of a cave is just another manifestation of this strange, aesthetic appreciation that we humans have of the world around us. We can obviously gain an insight into peoples' conceptualization of art thousands of years ago from these paintings and engravings decorating cave walls. But to gain an insight into when they first began to obtain some intellectual satisfaction from an appreciation of patterns and shapes in nature, we have to search elsewhere for clues.

People have been scribbling on rocks for a very long time. Many of these scratchings and engravings appear to be rather abstract in their style, and it is not easy to think of them as art. The earliest known evidence to date for material that may have been used for decorative purposes is pieces of red, hematitic ochre that may well have been obtained from deposits dated at about 165,000 years old in what is now South Africa.[3] Other examples from Paleolithic deposits in South Africa and dated at about 77,000 years old consist of two pieces of engraved red ochre. Both are scored with a series of parallel crosshatched scratches. Such abstract designs have been recovered from younger deposits in Eurasia and have been interpreted by some as having been created with symbolic intent.[4] Others have cautioned against ascribing a symbolic, nonutilitarian function to such etchings, suggesting, on the contrary, that they might provide evidence for a utilitarian recording or counting function.[5]

But it's not only on rocks that Paleolithic people expressed their desire to leave a mark for posterity. They also used fossils. About thirty-three thousand years ago somebody scratched with utmost care a very neat cross on a chalky white fossil nummulite. For protozoans (single-celled organisms that teem in our seas today in the same way they have for hundreds of millions of years), nummulites are huge, about the size of a dime. This particular artifact was found in Tata, Hungary, in rock shelter deposits containing what are known as late Mousterian[6] implements made by Neanderthals.[7] Although younger than the spectacular paintings at Lascaux and Chauvet, more poignant, perhaps, is a fossil brachiopod shell found in a fifteen-thousand-year-old Upper Paleolithic (Magdalenian) site in France. On its surface is engraved a human face. Could this, as Kenneth Oakley has argued, have symbolized a human soul to the person who drew it?

In France seventeen thousand years ago, someone scratched a spiral on a reindeer horn. At much the same time someone else engraved a heart shape on a rock, while another expressed his or her artistic ability by engraving a fernlike shape. Archaeologist John Feliks has suggested that these and many other similar seemingly abstract figures were inspired by natural shapes that the artists saw often in the world around them—in particular, the remains of fossilized animals and plants.[8] One of the more remarkable fossils found in these Magdalenian-age sites is an Ordovician trilobite that had been perforated with a pair of holes, presumably so that it could be hung on a thread. It was found in the nineteenth century in a rock shelter at Arcy-sur Cure in France and gave the site its name: La Grotte du Trilobite.[9] From the same layer in which the trilobite was found, a pendant was discovered, carved from a piece of lignite into the likeness of a trilobite-like beetle. Could the fossil have inspired the sculptor? This, too, had been perforated twice. Oakley, commenting on these finds, was convinced that both had amuletic significance.[10]

In many places fossils occur in the rocks that these people would have been using for many of their everyday needs. The very creation of a stone tool involved peering closely at the rock. It would have been hard not to notice these strange, sometimes abstract, at other times bilaterally or radially symmetrical shapes that were embedded in the newly created artifact. Given humans' propensity for artistic expression, it is not surprising that these strange shapes in the rocks could have inspired attempts to imitate them.

Five thousand years ago in Ireland someone engraved a circle in a rock. Around it he or she drew a spiral. And within the circle some lines were scratched that radiated out from the center of the circle.[11] What this meant to the artist we can never know. Neither can we tell if he or she was inspired by something from the natural world, or whether subconsciously this was an expression of some innermost thought or feeling. Either way it was a manifestation, perhaps, of the artist's inner self, consciously expressed in a manner that can be construed as being artistic, or creative. It is not hard to make the intellectual leap from a circle drawn around radiating lines—an act of creative expression—to the circular outline of a sea urchin, with its five equally spaced radiating lines, that lay in the center of a piece of flint. And would the people living on the chalk lands of southern England and northern Europe who thousands, or even hundreds of thousands, of years ago picked up fossil sea urchins be expressing any less a level of cognitive development, or creativity, than the person who inscribed a similar shape on a rock? After all, manipulating symbols is symptomatic of modern human activity.

And which triggered which? Did our ancient ancestors possess an innate facility to draw symmetrical shapes, which subsequently influenced their powers of observation of similar symmetry in the natural world? Or could the symmetry of nature, as reflected in fossils, have been the very inspiration for the dawn of artistic creativity? Could the casual observation of symmetrical shapes, like a circular fossil seared with a five-pointed star, actually have played a part in triggering aesthetic creativity in humans?

Ever since humankind attained its ultimate cognitive powers, the natural world with which early humans were so inexorably linked was a source not only of sustenance but also of objects with which the environment could be manipulated. When, a few million years ago, someone first picked up a rock and belted the living daylights out of some poor unsuspecting rat, humans' acquisition of natural objects began. While this distant ancestor might have failed to make a dent in the rat with one rock at hand, another might have done the trick. At some point in human development one of these ancestors may well have thought, *Mmmmm. That's a good rock. It killed my dinner. I'll hang on to that one.* It was then, through cognitive leaps such as this, that our propensity to collect, acquire, and accumulate took its first roots. No appreciation of the finer points of the color or shape of the rock was involved, perhaps, just the evolution of the concept of luck and, indeed, of conferring a symbolic significance to the object. Then why not hunt for another that looks the same, and so might just be imbued with the same quotient of luck?

The time when particular natural objects came to be collected for something other than purely practical reasons marks a critical period in the evolution of the human mind. To some people this represents the evolution of art, a uniquely human trait. Oakley suggested that fossils may well have been accumulated initially as items of curiosity or luck. Over time, they became transformed into objects that attained the status of items of greater symbolism, then perhaps into objects of magical and ultimately spiritual significance.[12] But at what point did they pass from being just imbued with luck to being awarded some more mystical form of power? Their believers might have felt that the fossils could alter the course of fate—even of their destiny. Could our ancient Swanscombe individual, who was so attracted to a fossil urchin four hundred thousand years ago, have had a sufficient level of cognitive development to afford the fossil some degree of mystical power in addition to being merely attracted to the fossil for its own sake? Could he and his fellow collectors then have had the ability to plot and plan how that power could be used to their advantage? And what can the meager archaeological record tell us of this transition, from lucky object to one perceived to possess supernatural power?

As I have indicated, it could be argued that the presence of the fossil urchin in the Swanscombe Acheulian tool was purely fortuitous—a chance happening, a one in a million event. After all, probably millions of such implements were made during the hundreds of thousands of years when generation after generation of *Homo erectus* and *Homo heidelbergensis* were churning out the same sort of tool. That being the case, the chance of a flint being used as a tool that just happened, inadvertently, to contain a fossil was likely to occur every so often. But, as I have argued, I believe there is sufficient evidence that points to the manufacturer of this particular hand axe having consciously selected this piece of flint solely because of the presence of the urchin. And I think that the feature on this fossil that took his fancy was almost certainly the striking pattern of a five-pointed star that was engraved on its surface.

So, is it possible that our lone soul, who some four hundred thousand years earlier had so assiduously collected his star-crossed stone, and who belonged in all probability to another species, evolved a mind capable of conceptualizing that had begun to approach ours? While he may have struggled to compose a light opera, had he, perhaps, attained a level of cognitive development that enabled him to be attracted so much to this shape that he wanted to keep the rock into which it was set for something other than a purely utilitarian function? And at what point in the evolution of human consciousness could an object like a fossil sea urchin have passed from being just such an object of aesthetic curiosity to one that possessed, in the mind of the finder, something more—something imbued with power?

> Plumbing the depths of the prehistoric mind is almost more difficult than solving the mysteries of the universe.
>
> **KENNETH OAKLEY,** *Decorative and Symbolic Uses of Fossils* (1985)

 2

First Collectors

It seems likely that the first collection of a natural object for a purpose other than to eat it was made for purely utilitarian motives. As humans' tool technology increased in sophistication and variety, so the scouring of the ground for suitable pieces of rock would have taken up increasing amounts of time. And in certain regions many of these rocks would have contained attractive symmetrical objects in the form of fossils. In addition to the star-struck urchins there were elegant scallop shells, spiraling marine snails, ammonites coiled like sleeping snakes, delicate corals, vase-shaped sponges, and many more. All such fossils have turned up in archaeological sites as objects specifically collected not only by farmers of Neolithic and later times (the last eleven thousand years in the Near East—just the past five thousand years in northern Europe), but also by earlier Mesolithic and Paleolithic peoples—those who hunted and gathered their food. For hundreds of thousands of years people must have sometimes gathered fossils, seeing them merely as objects of attraction rather than as objects of gastronomic utility—food for the mind rather than food for the body.

What these early fossil collectors called their fossils we obviously have no idea. Nor can we ever hope to know for certain what the maker of the

Swanscombe hand axe four hundred thousand years ago would have made of the five-rayed star on the fossil embedded in his piece of flint. It is hard to imagine that he would have seen much resemblance between the pattern on the fossil, which we now call a star, and the points of light that speckled his night sky, that we also call stars. Yet the reason why the two attracted the same name lies, I believe, at the root of why, for hundreds of thousands of years, people from many cultures across Europe, Asia, and parts of Africa collected and at times revered these strange little fossils.

The teardrop shape of the Swanscombe flint axe leaves little doubt that it is an Acheulian tool. Almost by definition, such hand axes are not only of the shape and size of this fossil-bearing one, but they are worked equally on both sides. Yet the hand axe bearing the simple urchin was only worked on one side by its manufacturer. Just three deft blows created a sharp edge on one side. But it was the largest, and perhaps the last, of these blows which points to the urchin being an object of singular significance, for it removed not only a good slice of flint but also took with it a sliver of the edge of the urchin. Had the toolmaker continued the usual Acheulian policy of knapping along the opposite edge to the same extent, he would almost certainly have removed a much greater part of the fossil. But apart from the tip of the axe, no attempt was made to work this other side. Although circumstance must not be immediately discounted, it does point to him treasuring the fossil in some special way. Moreover, it would imply that he possessed the necessary cognitive ability to reason that if he continued to modify the flint, he would destroy the very heart of the stone.

There may also be another reason, suggested to me by Gary Brown, assistant curator in Archaeology at the Liverpool Museum. When I visited the museum to look at the Swanscombe axe, I didn't realize that its current resting place would be in a former warehouse in the suburb of Bootle, the museum collections now being separated from the main museum. When we had located the drawer within which were resting scores of Acheulian axes, this one stood out, thanks to its distinctive urchin logo. "Here," said Gary. "Just hold it in your hand with your thumb resting in the groove in the fossil." Instinctively my index finger wrapped around the smooth, unworked side of the flint; the fossil nestled comfortably into the base of my thumb—it had become a razor-sharp extension of my hand. "Maybe placing his thumb against the fossil made him feel he was drawing on the power he thought the fossil had," Gary suggested. It would be a gross understatement to say that holding this axe in all probability in exactly the same way as the one who made it nearly half a million years ago was something special. There can be few other ways in which it is possible to feel in such close contact with someone who lived such an unimaginably long time ago. Holding his special axe also meant that I could understand a little more about him. He must have been quite large. My hands are pretty big and I'm nearly six and a half feet (2 m) tall; yet they curled around the axe perfectly. He would also have been right-handed.

This is not the only fossil that was picked up by people in England some hundreds of thousands of years ago, and probably for a similar reason. A fossil was found in a piece of flint in what is now the county of Norfolk, near a village called West Tofts; it was emblazoned in the center of one side of an Acheulian hand axe.[1] The fossil, an impression of the fossil bivalve *Spondylus spinosus*, is preserved on a weathered part of the flint.[2] In this case, the tool was carefully knapped to ensure that both sides were sharpened, leaving the fossil proudly displayed in the middle. It was almost as if advertising agents practicing for their submission to Shell Oil Co. got an inordinately early start on their competitors. But more than that, the fossil lies on a pedestal created by someone's deft craftsmanship, hundreds of thousands of years ago. Simple photos of the axe do not do the toolmaker justice. Not only does it sit proud of the rest of the axe, but the tool was crafted with such precision that the fossil sits geometrically in the center.[3] Only when the axe rests neatly in the hand can you appreciate the immense care that had been taken to make the fossil the focal point of the axe, indicating the existence of a profound aesthetic sense at this time.[4]

Holding these ancient artifacts in the palm of your hand today is an indescribably eerie feeling. You feel as if you are making a direct connection with the thoughts and feelings of someone else who handled them, but who lived hundreds of thousands of years ago. Loren Eiseley expressed this with more eloquence than I can ever hope to muster writing:

> Now the stone antedated anything that the historians would call art; it had been shaped many hundreds of thousands of years ago by men whose faces would frighten us if they sat among us today. Out of old habit, since I like the feel of worked flint, I picked it up and hefted it as I groped for words over this difficult matter of the growing rift between science and art . . .
>
> The mind which had shaped this artifact knew its precise purpose . . . The creature's mind had solved the question of the best form of the implement and how it could be manipulated most effectively. In its day and time this hand ax was as grand an intellectual achievement as a rocket.
>
> As a scientist my admiration went out to that unidentified workman. How he must have laboured to understand the forces involved in the fracturing of flint, and all that involved practical survival in his world. My

uncalloused twentieth-century hand caressed the yellow stone lovingly. It was then that I made a remarkable discovery . . . As I clasped and unclasped the stone, running my fingers down its edges, I began to perceive the ghostly emanations from a long-vanished mind, the kind of mind which, once having shaped an object of any sort, leaves an individual trace behind it which speaks to others across the barriers of time and language. It was not the practical experimental aspect of this mind that startled me, but rather that the fellow had wasted time.

In an incalculably brutish and dangerous world he had both shaped an instrument of practical application and then, with a virtuoso's elegance, proceeded to embellish his product. He had not been content to produce a plain, utilitarian implement. In some wistful, inarticulate way, in the grip of the dim aesthetic feelings which are one of the marks of man . . . this archaic creature had lingered over his handiwork.[5]

Transformation of unsullied nature into an item of "culture" was aptly expressed in culinary terms by the French anthropologist Claude Lévi-Strauss.[6] He spoke of turning the "raw" into the "cooked." Embellishment, for example by taking extra care and extra time on a tool with a fossil, adds the extra dimension of "gourmet": the "raw" flint is "cooked" into an axe or scraper, the gourmet touch being the intentional placement of the fossil in a central place on the tool, like an elegantly placed garnish.

But just what was it about the urchin or the shells that attracted these Paleolithic "chefs" in the first place? Was it the overall shape and symmetry of the fossils? Or was it something about their more detailed patterns? Help with answering these questions comes from other fossils that have been found associated with Acheulian material, often having been transported far from their original source by human activity. Of these fossils collected by our ancestral species, there is one in particular that indicates it may have been the starlike pattern that was attracting these early fossil collectors. This is what dominates the fossil urchin on the Swanscombe hand axe. On a similar, much smaller, scale it is found in a fossil that must have been transported by someone about four hundred thousand years ago.

In the same gravels at Swanscombe that yielded the urchin in the hand axe, two pieces of chert composed of the Late Jurassic coral *Isastraea oblonga* were found.[7] Within each of the tiny corallites (skeletons of coral polyps) that make up the coral colony are radiating, starlike struts that in life supported the soft coral animal. No outcrop of rock containing these fossil corals occurs anywhere in the Lower Thames valley in which the gravels were deposited. Both pieces are humanly struck flakes, providing little doubt that the fossils were brought to the site, presumably by *Homo heidelbergensis*, more than a sixty miles (100 km) away, perhaps from the Dorset coast, where these fossil are known to occur. Given that this chert is not as good as the local flint for tool making, it seems likely that it was the pattern—a constellation of little stars—that attracted the finder.

Other worked tools containing fossil sea urchins are known from these ancient times. However, they were not crafted into typical Acheulian hand axes. One, found in Pleistocene river gravels in association with Acheulian artifacts at Saint-Just-des-Marais in France, is a most unusual specimen of the fossil urchin *Micraster*.[8] Shaped like a heart, with a small five-pointed star at its center, the entire fossil has been transformed into an artifact. The *Micraster* is a water-worn flint cast, the periphery of which has been

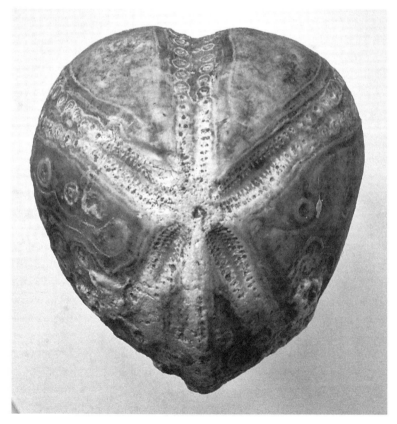

Fossil urchin *Micraster cortestudinarium* collected by Mr. Alan Smith and known by him and his wife, Betty, as a "sheep's heart"; from a field in Linkenholt, Hampshire, and kept by the front door of his cottage. Fossil length 1.8 inches (4.7 cm). Author photo.

skillfully chipped, to transform the urchin into a useful scraper. It has been most carefully worked to ensure minimal damage to the specimen, to ensure that the apex of the fossil lies exactly at the center, and from this the five worn rays radiate—a notable gourmet embellishment.

Francois Poplin of the Muséum national d'histoire naturelle in Paris studied a similar specimen from La Roche-au-Loup.[9] He has argued that objects like this are crucial to our understanding of the development of various aspects of conceptual thought in human evolution. In Poplin's opinion, though, it was the symmetrical shape of the urchin, with its radial symmetry, that appealed most to the collector, rather than the five-pointed star as such. Such radial symmetry, he argues, is indicative of an underlying simplistic and generalist view of the world. During the evolution of human cognitive development there was a transition from such simple, symmetrically enhanced scrapers to the next stage, which was the typical middle Paleolithic bilaterally symmetrical hand axe. The last stage in this evolution, Poplin suggests, is the development of the lateralized tool in which symmetry has broken down, like the knives we use today to slice our roast lamb or butter our toast.

It is a little ironic to think that the Swanscombe individual who treasured his hand axe decorated by an urchin unwittingly imparted elements of all three of these stages: the rhythmic balance of the overall bilateral shape of the hand axe and the radial form of the urchin at the center, offset by the incomplete working of this tool. The result was its transformation into a lateralized implement that presages what was to evolve as a more conscious exercise hundreds of thousands of years later. It also looks as though another species of *Homo* was also not averse to collecting the odd fossil sea urchin—*Homo neanderthalensis*. A flint that was used as a tool has been found at Tercis-les-Bains in Landes in southwest France.[10] Fashioned in the Mousterian style typical of Neanderthal technology, the tool has a specimen of the fossil sea urchin *Cyclaster* tucked into one side of the flint. It had been worked in such a way as to leave the cutting edge facing away from the fossil. The result was to enhance the appearance of the urchin on the flint.

Until very recently, this was the sum total of known worked Paleolithic flints within which were encased fossil sea urchins. The reason for this paucity of examples probably lies in the fact that these were all chance finds. Nobody had set out specifically to search for such objects. But over the last ten years the fascination of University of Rennes paleontologist Didier Néraudeau for early people's fossil collecting habits led him to excavate some deposits in western France near the mouth of the Charente

Mousterian flint scraper with fossil sea urchin made by *Homo neanderthalensis*, Charente River, France. Bar represents 0.8 inch (2 cm). Photo courtesy of Didier Néraudeau.

River. These had long been known to yield Paleolithic flints. What he discovered exceeded his wildest dreams. Of the hundreds of Acheulian and Mousterian types of tools that were uncovered, twelve contained fossils.[11] Of these, two are brachiopods, one is an oyster, and one a star-shaped crinoid stem plate, while the remaining eight are sea urchins. Some of the urchin fossils are of round urchins, others are parts of heart-shaped urchins, while some are impressions of large tubercles to which large spines were attached on the urchin.

These fossil sea urchins are present on both Acheulian axes and Mousterian scrapers, suggesting that members of both *Homo heidelbergensis* and *Homo neanderthalensis* had a hand in collecting these fossils. In one Acheulian axe the fossil juts prominently from the base; as well as being decorative, it would have made a useful handle. In another, the five-pointed star of a sea urchin called *Bolbaster* is splayed uncompromisingly on the side of a Mousterian scraper. When grasped in the hand the star would have rested firmly against the user's palm. One of the more intriguing flints bears the impression of a large tubercle. It looks for all the world like a large eye staring out with a frozen, flinty gaze from the depths of time. Given that these patterns were naturally engraved on the flints that these early hominids were finding, is it any wonder that, on a rock more forgiving than this hard flint, someone had first tried to imitate such compelling images by scratching a few tentative lines and circles? Is this when art was born? While *Homo heidelbergensis* and *Homo neanderthalensis* might have appreciated the natural symmetry of these shapes, did it inspire them to try and imitate it, or was that confined to the third species of *Homo* who also took a fancy to fossils—*Homo sapiens*?

The explosion of artistic expression exhibited by *Homo sapiens* that occurred during late Paleolithic times (35,000–10,000 years ago) coincided with the final surge in intensity of the last Ice Age. During this period much of Europe, when not covered by ice, was cold tundra. Mammoths, woolly rhinoceroses, and reindeer shared the cold grass steppes with the early form of *Homo sapiens*, Cro-Magnons, who hunted these large mammals for food, clothing, and bone tools. For brief periods the ice relaxed its grip, and forests returned to the region: pine, hazel, lime, elm, and beech, with willows weeping along the watercourses formed by the melting of the ice and permafrost. During these warmer times, the mammoth, rhinos, and reindeer were replaced by bison and horses—but not horses that we would recognize today. These animals were long-headed, massive of limb, and broad of hoof. But at times these Cro-Magnon hunters themselves became the hunted, and the prey of bears, lions, and hyenas.

In the more sheltered regions of southern Europe these people found time, when not hunting, to daub the walls of their caves, such as those at Lascaux and Chauvet in France and at Altamira in Spain, with spectacular galleries of animals. These are the earliest animal paintings of which we are aware—paintings created up to thirty-three thousand years ago. Rendered in red and black, these paintings were faithful depictions of the creatures the Cro-Magnons hunted and with which they shared their world. Up to forty thousand years ago, some of these people even engraved and sculpted blocks of stone that had fallen from the roofs of their caves with depictions of animals.[12] Their propensity for aesthetic expression also extended to their own bodies. Whether they painted or tattooed themselves we do not know. But we do know that they perforated shells and mammal teeth, such as those from deer, and even fish vertebrae, and then strung them together to wear as necklaces and bracelets. And they collected fossils that they also wore as personal adornments. One of only four movable cultural objects found at Lascaux was a fossil gastropod shell that had been carefully sawed to produce a slit that could be used for threading.[13]

Fossil urchins, too, were drilled and used as jewelry. These fossils, though, were not the heart-shaped urchins like *Micraster*, but flattened spheres that looked like hard little doughnuts when drilled. More than eighty of these urchins from a dozen sites in southwest France have been found.[14] Some were perforated as much as thirty-five thousand years ago. Sometimes they were used in combination with other items. A necklace from late Perigordian times (that's about twenty thousand years ago) discovered at La Gravette had a fossil urchin as its centerpiece, flanked with mammal teeth and perforated shells. At a site called La Placard in Charente, a necklace of fifty-nine perforated fossil urchins of the species *Sismondia occitana* has been found. These may once have been threaded onto a necklace.[15] These examples represent some of the first evidence we have that our species, like *Homo heidelbergensis* and *Homo neanderthalensis*, also liked to collect this type of fossil.

At this time, when the world was usually in the fierce grip of a continual winter, people elsewhere in Europe were also adorning themselves with other parts of fossil urchins. During excavations carried out in Dolní Vestonice in the Czech Republic in the later 1930s, fossil brachiopods and sea urchin spines were located in the dwelling places of the so-called Gravettian mammoth hunters.[16] It is possible that the spines were collected merely as items of intrinsic interest. Yet we know that these late Paleolithic people enjoyed covering their clothes with all sorts of naturally occurring objects, such as shells, teeth, and beads made from ivory, limestone, jet, and bone. This has been shown by the discovery of skeletal remains in Russia that were swathed in ivory beads and polar fox teeth. One small child was clothed in more than five thousand beads.[17] The Gravettian hunters are known to have been avid collectors of all sorts of fossils. Some, like the narrow, straight shells called scaphopods, both recent and fossil, they used in necklaces. From the 35,000-year-old site of Kostenski on the Dom River in the Ukraine, perforated pieces of belemnite[18] have been uncovered.[19] Gravettian hunters also made beads from fragments of belemnites, drilling tiny holes through them to enable them to be threaded onto necklaces. One such fossil was found that had been polished and perforated with a hole at one end, to produce a beautiful translucent pendant.[20] Kenneth Oakley was of the opinion that the rich amber color of these fossils may have played a part in inspiring their wearers to believe that the pendants possessed magical powers.[21]

The fossil sea urchin spines may also have served a similar decorative purpose. Unlike the belemnite beads they are not perforated in any way, implying that they were not used in necklaces or bracelets, but they may well have been attached to fur clothing. Interest by late Paleolithic people in fossil sea urchin spines extends south at least as far as Lebanon. Two club-shaped spines of the cidaroid urchin *Balanocidaris glandifera* were found in late Paleolithic levels at the Ksar 'Akil rock shelter in that country during excavations immediately prior to and following the Second World War.[22] In more recent times such spines are known to have been used as talismans for over three thousand years.[23] Called lapis judaicus, or jew stones, because they were often found in Judea, their frequently swollen shape led some of them to be used for treating ailments of the bladder. According to physician, natural historian, and antiquary John Woodward (1665/1668[24]–1728),

> The bodies called *Tecolithi* by Pliny, *Lepidus Judaici* and *Syriaci* by other writers, so much celebrated by the ancient Physicians for their diurectic properties but reputed by all as no other than stones, have been at last publickly demonstrated to be only elevated spikes of *Echinus ovarii*, brought forth of the sea at the Deluge.[25]

The first reference to "jew stones" having a medicinal use was in the *Materia Medica* written by the Greek Pedanius Dioscorides, who lived in the first century AD.[26] His recipe was first translated into English by John Goodyer in 1655:

> But ye Judaicall stone grows in Judea, in fashion like a Glans, white, of very handsome form, having also lines answering one another as if made by turning. Being dissolved, it yields no relish in ye taste. But a Cicer-like bigness (thereof) being dissolved like a Collyrie on a whetstone with three Cyathi of warm water & drank, is of force to help Dysuries & to break ye stones in ye bladder.[27]

Others considered that the stone's efficacy was enhanced with a concoction of other ingredients.[28] A typical recipe of the seventeenth century was given by Gualtherus Bruele in his *Praxis medicinae*, published in 1632:

> A ataplasme of the flowers of Cammomill, Mellote, meale of Linseed, Fengreeke, Wheate and Lupines. *Lythontripon*, with Turpentine, washed,

or with the decoction of the roots of grasse and Fennell. *Elect. Iustinum*, of the ashes of Scorpions, *Dialacca*. The decoction of the wood *Guaiacum*. The ashes of a Scorpion, the iaw of a Pike, Egshelles when the Chickens are hatched. The great conserue of our description is a good preseruatiue against the stone. Conserue of the roots of Parsley, Radish, water-Cresses, Turpentine burnt vpon a hote tile, the stones of Peaches, and Cherries. Hot Goats-blood, that it may not curdie in the belly, a little of the *Coagulum*, or seed of the Goate must be added. The powder of a hare with the skin dried in the fornace; the fruits of *Alkekengi* bruised, and strained, when they haue beene steeped in wine. *Lapis Iudaicus*. A Wren, a Wagtaile, Sampier, seeds of Saxifrage, Burnet, *Ruscus*, Fennell, Parsly, Radish, *Milii solaris*. Broome. A decoction of Radish, red Chiches, Parsley, *Ruscus*, Sperage, Mallowes, Fennell, Nettle, Saxifrage, roots of Grasse. The rindes of sharpe Radishes, bruised, and macerated in white wine. The flowers of Broome, macerated in Cammomill, and so set in the sunne. oyle of scorpions, these may be vsed when Cherries, & oyle of medicines that doe break the stone are giuen.[29]

Christopher Wirtzung, writing in his *General Practise of Physicke* in 1617, described a recipe that not only served to provide relief against bladder stones, but at the same time guarded against disease. As Chris Duffin points out in his review of the use of fossils as drugs, the prophylactic nature of this recipe was perhaps brought about by the addition of the wing covers of the beetle *Cantharides vesicatora*, otherwise known as spanish fly.[30] Other objects added to the mix along with the "jew stone" included "ashes of burnt Scorpions," "Bucks blood," "Hares ashes, ashes of Wagtayles." A range of herbs including caraway, saxifrage, hollyhock seeds, pepper, mallows, maidenhair, and roses were added for good measure.[31] And even today these "jew stones" can still be bought for medicinal purposes in markets in Jordan, Israel, and Pakistan.[32]

These fossil sea urchin spines continued to be used for their perceived medicinal powers in Britain throughout the seventeenth century.[33] Importing such fossils was so common that during the time of Elizabeth I and Charles II an import tax of one shilling was imposed for every pound weight of the fossils.[34] Even though their true nature was known by the mid-nineteenth century, they were still being prescribed to help cure bladder ailments. In medieval times they were also used extensively throughout the Near East for helping to cure not only kidney stones, but also snakebites.[35] Their use as an antivenin was described by Moses Maimonides, a Jewish philosopher, physician, and jurist, writing in 1211. To pulverized *lapis judaicus* he recommended adding a range of other ingredients, including crocodile fat, pigeon and duck excrement, and goat dung, not forgetting the onion and honey. All was kneaded together into a plaster and added to the wounds as a poultice to draw out the poison.[36]

It has been argued that the appearance of beads and pendants during Paleolithic times marks an important milestone in the evolution of human behavior. The widespread use of such objects made from shell, teeth, ivory, and stone, including fossils, has been seen as the "hallmark of the upper Palaeolithic."[37] These artifacts have been interpreted as symbolic communication media that mark the evolutionary development of cognitive abilities reflecting "modern behavior," including language.[38] Yet the evidence from the early Paleolithic symbolic use of fossil sea urchins suggests that such a level of cognitive development had evolved much earlier.

It is intriguing to wonder just what Paleolithic hunter-gathers really thought about the fossils they were collecting, particularly the fossils covered by a five-pointed star. An interesting shape, whose symmetry in some way appealed to them, no doubt. But did they make more of it than that? An attraction, perhaps, to the five-rayed patterns, as well as an innate insatiability for collecting natural objects simply for the sake of it. But could they also have seen in this shape a mirror of their own form—arms and legs outstretched, like a little fossilized Vitruvian Man? (see chapter 10). They were certainly fascinated by fossil sea urchins to the point of having them adorn their tools, their bodies, and their clothes.

The information that we can derive from objects that have been altered by the activity of humans is quite different, though, from the information imparted by the more indirect association of fossilized urchins with human activity. One practice that made its appearance not long after the glaciers began their long, northward retreat, when our ancestors were preparing to give up their hunter-gatherer ways and settle down to a sedentary farming existence, was their habit of burying fossil urchins with their dead. This was not so much to adorn the bodies of the dear departed, but more likely a reflection of a spiritual belief that appears to have been embodied in these little pebbles marked with a five-pointed star. Some of our earliest ancestors to carry out this practice were Mesolithic people living more than six thousand years ago in the Scania region of southern Sweden. Two graves excavated at Skateholm each contained bodies that had been buried with fossil sea urchins placed carefully at their hips.[39] Fossil sea urchins have also been recovered from other Mesolithic sites in southern Norway.[40] Whether earlier Paleolithic peoples ever buried their dead with fossil sea urchins we simply do not know. However, the practice appears to have become relatively commonplace during Neolithic times, and through the succeeding Bronze and Iron ages. It also extended into Saxon times as "pagan" ideas intermingled with Christian beliefs, and souls were sometimes still sent on into the next life accompanied by fossil urchins. While occasionally other fossils were used, it is fossil urchins that were the most popular, especially in northern Europe.

So, unless we are to argue that these fossils just fell into the graves all by themselves, it seems that people were making a conscious decision to place them with the dead, arising from a perceived significance that was attached to the fossils. In all probability this significance was much greater than was attached to the crudely manufactured, much older flint tools containing embedded fossils. This practice of burying fossil urchins in graves may have started in a simple way during late Mesolithic times, with one fossil urchin per body, when people were acquiring a more settled way of life. But what drove these people to collect fossils in the first place? Why, of all things, place hundreds, even thousands, of fossil sea urchins in a grave? Had the need to satiate the mind evolved to satiate the soul?

To understand the motives of Paleolithic and Neolithic peoples we must go far back in human history, far beyond the first written records. As we travel so far back into the realm of memory, knowledge fades, and it becomes harder and harder to see more than just faint outlines of fragments of images. The great chain of information, along which knowledge passes from one generation to another, becomes more brittle the further we travel. Breaks begin to appear, and judgments have to be made of what these missing links meant. And as the journey continues deeper and deeper into time there are more gaps than links, until finally there is nothing left but a ghostly image of the oldest link and the faint sound of naked voices in the dark.

But this most ancient of links still persists, I believe, in the memories that have lingered and been passed from generation to generation over many thousands of years. Whether this link extends as far back as the *Homo heidelbergensis*, fascinated as he was by the fossil urchin that he picked up, with those of us still to this day who are just as fascinated with these objects—we who assuage our lust for collecting by searching for fossil sea urchins from fields and beaches, and from quarries and cliffs—is questionable. Even so, within this strange and curious passion that some still share for collecting these star-crossed stones, there lies an unutterably faint but cogent link that connects us to the lone wanderer who picked up a fossil urchin on the chalk downland nearly half a million years ago. And maybe there is even the faintest, almost imperceptible sigh of a societal memory that touches the person who wears a five-pointed star pendant, places a five-pointed star on top of a Christmas tree, or places stars around a window in a church—a memory from those people who enshrined these fossils in the graves of their loved ones thousands of years ago.

> The passion for collecting, which leads a man to be a systematic naturalist, a virtuoso or a miser, was very strong in me.
>
> **CHARLES DARWIN,** *The Autobiography of Charles Darwin 1809–1882*

 # 3

Urchins

Like many children living in the south of England in the 1950s, much of my life, when not spent kicking a soccer ball around the streets, seems to have been taken up with collecting something or other. While my very distant ancestor might have developed a preoccupation for collecting round flints with five-pointed stars on them, it really didn't matter what I collected: the "collecting bug" had taken up residence in my bones at an early age. The most sought-after treasures, however, were the little cards that were to be found nestling in packets of Brooke Bond PG Tips tea. Here was my highlight of the week: launching into the bag of groceries to grab the packet, and fumbling frantically with the top that always seemed to have had just too much glue put on. I was convinced that tea manufacturers employed the most miserable, inhuman person they could find to make it as hard as possible for me to open the packet. There he sat, at his dark brown, stained table in a dingy corner of the factory, slapping as much glue to the flap at the top of the packet as he possibly could, a satisfied leer twisting the corner of his thin mouth as he pushed the flap down with all his might. Another eager child frustrated.

I was in no doubt that the main goal of this drab person was to glue the packet together with such efficiency that as I tried to rip open my treasure box I would have to tear so hard that the whole packet would be rent asunder and the contents of the tea sent sailing over the floor, over the cat, and—to the ultimate triumph of the evil gluer—all over my mother. This time I was glad to see that the tea packet-gluer had failed. However, he had another card (so to speak) up his sleeve. Or in this case in the sleeve between the bag of tea and the packet. Fortunately, no doubt for the sake of hygiene, so as to inhibit small children from sticking their grubby little fingers into the loose, finely shredded leaves, the cards weren't immersed deep in the tea by our glue-slapping friend. They were cunningly hidden between the bag of tea and the box. Inevitably, of the four sides to choose from it always seemed to be the last that yielded up the treasured card.

But when the card was finally retrieved, what joy, what ecstasy—another card to cherish, to hoard away in a box. It really didn't matter what was on the card. While the sets of illustrations of Bird Paintings, British Wildlife, or British Wildflowers (never, unfortunately, British Fossils) no doubt sank somewhere (to some effect, I hope) deep into my subconscious, it was the pure and simple joy of possessing the cards, and of collecting more and more, that was the true delight. However, the frustration arose as quickly as the joy subsided with the awful realization that it was to be another week before another packet of tea could be ripped open and another card added to the collection.

So other things had to be collected. Down the hill to Mr. Tickner's sweet shop to spend a penny on a thin sheet of bright pink, sickly sweet-smelling bubblegum—and a card! The gum was soon dispensed with, but the cards were added to the ever-growing collection. More flags, or soccer teams, or cricket stars—who cared? What else to collect? Car numbers, train numbers, and autographs of soccer players and cricketers. But my collecting wasn't confined to man-made objects. I soon discovered a natural world brimming with myriad objects to collect: long trudges across the back of the South Downs in Sussex on which I lived, through the snow to skeletal hawthorn trees to collect abandoned birds' nests; boggy ponds in the heat of summer to dredge with a jam jar for tadpoles and newts, or even—most prized of all possessions—sticklebacks. All were collected in great triumph, but usually stuck in a jar of water that became fetid within a few days, leaving a collection of tadpoles that dwindled before my eyes as these ravenous, overgrown spermlike creatures became rampaging cannibals and my precious collection self-destructed.

Not surprisingly, I never bothered to consider why I collected all these things. It was just a thing I did, like eating or sleeping. After all, how many of the millions of people who are similarly addicted to collecting ever do ponder why they do it? We can always sit back and try to convince ourselves that our unique collection of seventeenth-century Bolivian handkerchiefs is the finest this side of Birmingham, and that it must be worth an absolute fortune; but there are few of us who would ever part with these collections, so lovingly accumulated, that have become every bit as much a part of our lives as the cat or dog, or even our husband or wife.

Looking back on my own fervent early days of fulfilling the collecting imperative, I had no idea that I was merely satisfying my species' innate curiosity and inquisitiveness, and I don't really think that I was in any way an emotionally deprived child, as Werner Muensterberger might argue—on the contrary. These cards, these autographs, these birds' nests were no parent substitute. This was something much deeper: an innate genetic imperative to go on the hunt, to capture whatever it was that my psyche demanded had to be collected. Yes, it was always good to have the collection, but on reflection in many ways it was the thrill of the hunt that drove me on.

Nor did I stop to think what lay beneath the crusty layers of housing estates that partly covered the flowing downland beyond my bedroom window. It was the ground, that's all I needed to know. That the ground was made of rocks never entered my head. That is, not until the evening when my mother returned home from a holiday in Switzerland. I was eight years old. I stood by the dining table desperate to know what she had brought back for me—a toy car? A cuckoo clock? But all she handed me was what seemed at first sight a most unprepossessing present. It was a small piece of gray rock, about an inch and a half long, and half an inch wide. *So what?* I thought, until she said, "It's a piece of the Alps." I was transfixed. Here I was, in a little room, in a small town in England, actually holding a piece of the Alps. Move over, Brooke Bond tea cards! Forget the newts and the birds' nests! Tip out the tadpoles and the sticklebacks! This was it: rocks, the ultimate treasures to collect. I would collect them from wherever I went, from every mountain range, every cliff and every hill.

As the lumps of rocks started to accumulate beneath my bed, it didn't take me long to realize that beneath my feet was not only rock, but objects within the rocks that I could also collect—fossils that represented the remains of animals and plants long since vanished from the Earth. The rock type upon which I lived, and which forms the South Downs, is chalk. It was deposited as a soft, white sludge in a warm sea 70 million to 80 million years ago, and was lifted from the seas to form rolling downs by the very same tectonic forces that, more spectacularly, formed the Alps. However, rather than being part of the major powerhouse that lifted the Alps skyward tens of thousands of feet, here, far from the action, the faint ripples of pressure had only been enough to gently propel this ancient seafloor just a few hundred feet above sea level. Aided and abetted by the erosive power of the sea to form towering white cliffs, as well as by quarrymen anxious to produce the raw material for cement, and farmers anxious to plow the chalk downland, it has been yielding fossils to collectors for hundreds of thousands of years.

Following the birth of the science of geology at the beginning of the nineteenth century, the first and most famous collector of fossils in the South Downs was Dr. Gideon Mantell. Together with his wife, he found some of the first recognizable dinosaur bones in Tilgate Forest in Sussex in the 1820s.[1] But Mantell's other claim to fame was the publication of his first great work documenting the fossils of the South Downs. Called, not surprisingly, *The Fossils of the South Downs*, this landmark study was based on extensive collections made from quarries and coastal cliffs in Sussex.[2] While Mantell collected many fossils himself, especially in his youth, as his medical practice expanded he also became a "gatherer" of fossils, as well as being a "hunter."[3] Those who hunted were the quarrymen, whose backbreaking work often yielded fossil treasures. These provided a useful source of income when sold to the likes of Dr. Mantell, who periodically stopped by to purchase the latest finds.[4] As I embarked on establishing my own collection of chalk fossils some 140 years later, I was unwittingly following in Mantell's footsteps: delving into the quarries that pockmarked the northern face of the Downs. These chalky treasure chests soon yielded many objects to slake my thirst for collecting fossils: smooth and ribbed brachiopods, knobbly sponges and elegant bivalves, and even, if I was lucky, a fish vertebra or shark tooth or two.

Sometimes the fossils were preserved in white calcite, like delicate, carefully crafted pieces of porcelain. Other times they were quite the opposite: hard, gray flint that originally had formed as layers or nodules within the chalk. The flint nodules peppering the chalk like raisins in a fruitcake were formed from silica gel that had developed from the dissolved remains of millions of sponges that must have carpeted this chalk seafloor. Gently permeating the inside of dead shells, or of sponges themselves, the portions of soft silica gel eventually hardened as the mud turned into chalk around them and they were transformed into beautiful flint molds.

The fossil that I began to seek more than any other seemed also to be the most unattainable, my own Holy Grail: a fossil sea urchin. What was it about this particular type of fossil that inspired in me this peculiar fascination? Perhaps it was just the fact that it was so different from anything else that I had ever seen. Or maybe it was the peculiar heart shape that many possessed. But more likely, on reflection, it was the five-pointed, starlike pattern that was etched onto its surface: this innate human attraction to the pentagram. Whatever the reasons, the result was that most weekends, whether the Sun shone or the rain fell and the chalk dust turned to white mud that clung to everything with the tenacity of a leech, I scoured the downland quarries for urchins. I can't say that I was ever very successful. But the feeling was always there that the next rock that I hit with my hammer just might contain my Holy Grail.

Gideon Mantell had more success than I ever did. Among the thousands of fossils that he had collected from the chalk deposits of southeast England were those of many sea urchins. The first three decades of the 1800s, when Mantell was collecting, was a period when the science of geology was in its infancy.[5] Mantell and others in England, like William Buckland, Charles Lyell, and Adam Sedgwick, all regarded fossils as the petrified remains of animals and plants. But less than two hundred years earlier, naturalists and philosophers had very different ideas. Were these "figured stones" once living creatures or, as most thought, were they just stones that bore a superficial and coincidental resemblance to now-living organisms? Country people, though, had no such difficulty—fossils such as the star-crossed stones were "thunderstones" or "shepherds' crowns," lucky objects to be carried around in your pocket or taken home, to protect you from evil.

Yet, surprisingly, 2,500 years earlier the early Greeks were in little doubt that fossils represented the petrified remains of once-living animals. Xenophanes of Colophon (540–510 BC), along with Pythagoras of Samos (540–510 BC) and Xanthus of Sardos (flourished 480 BC), observed marine shells in mountainous regions and pointed out that this meant the area had once been under the sea.[6] Similarly, the historian Herodotus came to the same conclusions based on marine shells he found inland in Egypt. He was less clear, however, on fossils of shelled protozoa nummulites (foraminifers) that are common in limestones of the pyramids, believing that they were the remains of lentils left over from food brought for the pyramid builders. However, the commonsense views of these Greek writers were overwhelmed for the next 1,500 years by a mixture of fantastic and mystical explanations.[7] Theophrastus (368–284 BC), for instance, considered fossil bones to have been formed by a "plastic virtue latent in the earth,"[8] a view that was to persist for the next two millennia. Pliny the Elder (AD 23–79), in his *Natural History*, which formed the basis for understanding the natural world until well into the Renaissance era in Europe, thought the resemblance of fossils to living animals was mere coincidence. Thus, what we now recognize as the tooth of a fossil shark was called Glossoptera by Pliny, who thought that it "resembleth a man's tongue, and groweth not upon the ground, but in the eclipse of the moon falleth from heaven, and is thought by the magicians to be very necessary for panders and those that court fair women."[9]

Apart from individuals such as the Muslim Aristotelian philosopher Avicenna (980–1037) and the German Albertus Magnus (1193?–1280), who accepted the organic nature of fossils, most other writers throughout the Middle Ages believed that fossils were formed by a "plastic force" or "formative virtue" that existed within the Earth. This *vis plastica* was forever trying desperately to form creatures and plants, with fossils being the unsuccessful products. One of the first to adopt a more rational approach to the origin of fossils was the Swiss naturalist Conrad Gesner.

Shortly before he died of the plague in 1565, Gesner's *De Rerum fossilium, Lapidum et Gemmarum maxime, figuris et similitudinibus Liber* (A Book on Fossil Objects, Chiefly Stones and Gems, Their Shapes and Appearances) was published. It is a small book and contains a mere 169 pages. But what it lacked in size it more than compensated for in the impact that it had on natural science. The historian of paleontology Martin Rudwick has argued that this book, written by the greatest naturalist of the sixteenth century, played a pivotal role in the emergence of paleontology as a science.[10] What made Gesner's book so significant was his recognition that some objects found in rocks and known as "fossils" were the remains of once-living organisms. The "figured stones" that Gesner presented as a series of woodcuts represent the first time that fossils had been illustrated in any book in such a comprehensive manner. Among them were forms clearly recognizable as sea urchins, some species still in existence, some extinct. Crucially, Gesner accurately identified a flint fossil as a "petrified" sea urchin. However, he failed to recognize another fossil studded with knobbly tubercles as a fossil urchin.

Today we can clearly identify Gesner's excellent woodcut as a specimen of a cidaroid sea urchin. But he was not familiar with this type of urchin, so he maintained the traditional view that it was an *ovum anguinum*, or serpent's egg (see chapter 12). Despite being the most renowned naturalist of his age, and recognizing the organismal affinities of some of the fossils, even Gesner found it hard to shake off entirely the cloak of tradition that clung to many of the fossils. His contemporary Michele Mercati (1514–1600), in his *Metallotheca Vaticana* (1574), also illustrated similar fossil sea urchins, and he, too, called them *ovum anguinum*.

For its time, Gesner's suggestion that some of these figured stones had once been living animals was decidedly radical. But he was not entirely alone. Frenchman Bernard Palissy (1510–1590) was a journeyman potter always on the lookout for material he could use in his ceramic making. Regarded by some as the father of French ceramics,[11] Palissy also became a scientist, land surveyor, religious reformer, garden designer, glassblower, painter, chemist, geologist, philosopher, and writer. He frequently depicted marine animals on his ceramic works, and also became interested in fossils. Self-taught, and knowing neither Latin nor Greek, Palissy took great delight in thumbing his nose at the "learned" writers of the time. He was also decidedly down-to-earth in his explanations of natural phenomena, accepting neither the traditional view that the biblical Flood explained all geological phenomena, nor the prevailing ideas on the origin of fossils. In his *Discours admirables de la nature des eaux* (1580), Palissy recalled how "a lawyer showed me two stones exactly similar to the form of the shell of the sea-hedgehog [urchin]. The said lawyer believed the stones to have been shaped by a mason and was much surprised when I gave him to understand that they were natural, for I had already judged them to be shells of sea-hedgehogs which in the course of time had been turned to stone." These radical ideas played a part in Palissy's undoing and his becoming the first martyr to paleontology.[12] Not only did he take up with the Protestant Huguenots, but after the publication of his book he began lecturing on his ideas about fossils and petrified animals. The result was that, despite the best efforts of one of his patrons, the queen, Catherine de' Medici, he was charged with heresy and thrown into the Bastille. Given that he was at that time seventy-eight years old, it is not surprising that after two years he succumbed to malnutrition and consumption, becoming the first victim of persecution in the history of paleontology.

The realization that fossils such as sea urchins are of organic origin was also dawning in Italy. Naturalists in this country were foremost in the demolition of the medieval ideas about fossils and began to adopt a more rational approach to understanding their true nature. One of the first to describe fossil urchins was the architect Leone Battista Alberti (1404–1472), but he had no idea what they were. But following Leonardo da Vinci's unpublished views plainly arguing for the organic nature of fossils, Girolamo Fracastoro (1483–1553), in an interview given in 1517 but not published until 1622, observed,

> When our distinguished citizen Torellus Sarayna the lawyer . . . was describing that part of the mountain of our city [Verona] where there is a

chalybeate spring, and the caverns and hollows found in that region when the Venetian signory was building the fortifications . . . another thing seemed to me a miracle, that when the mountain was being dug, there were seen sea-urchins of stone, crabs, sea-snails, cockles, oysters, starfish, birds' beaks, and many other things of that sort.[13]

Likewise, in his *Historia Naturale* (1599), Ferrante Imperato (1550–1625) figured some excellent illustrations of fossil sea urchins, and thought they were formed "not by their own vegetative power, but from preceding forms of the remains of marine animals, which by the decay of the ligaments were deprived of their spines and membranes."[14] The conflict between those who took a more rational approach to the origin of fossils and those who did not is perhaps best exemplified by the Sicilian artist and naturalist Agostino Scilla (1629–1700), in his book on fossils entitled *La Vana Speculazione disingannata dal Senso* (Vain Speculation Undeceived by Sense), published in 1670. The frontispiece shows "Sense" holding a recent sea urchin under the nose of the unbelieving, wraithlike "Vain Speculation" while pointing to fossils lying on the ground, including that of a sea urchin, and pleading for "Vain Speculation" to forget about superstitious ideas and accept the obvious, organic nature of fossils.[15]

Elsewhere in the Mediterranean region at this time, however, superstitions concerning the remains of fossil urchins still were ingrained in folk culture. During an extraordinary sermon given in Malta, the Augustinian monk Padre Spirito Angusciola in 1566 ascribed many of the fossils commonly found in the local limestone as evidence that Saint Paul had once trod the island. Some fossils were described as Lingue (Saint Paul's tongue), others as Mammelle (his breast), and some as Bastonicino (a part of Saint Paul's anatomy I will discuss below). Collectively they are known today as the Vestigie di San Paolo[16]—vestiges of the saint's anatomy embedded in the rocks of Malta as a reminder of his visit to the island. While the Lingue refer to fossil shark teeth, both the Mammelle and Bastonicino we now know to be the remains of fossil urchins.

While Scilla, in his *De corporibus marinis lapidiscentibus* (1752), figured plates from fossil cidaroid urchins that bear prominent tubercles which he described as nipples, he failed to ascribe them to Saint Paul. But earlier, Paolo Boccone (1633–1704), in his *Rechercches et Observations Naturelles* (1674), specifically called them Mammelle di San Paolo. The spines of cidaroid urchins, common in the Tertiary limestones of Malta, were considered by some as representing the Baculum, or staff, of Paul. Scilla observed that "In Melita vulgo appellantur Baculi S. Pauli,"[17] even though he

Frontispiece from Agostino Scilla's *La Vana Speculazione disingannata dal Senso* (1670), showing "Sense" trying to convince "Vain Speculation" that the fossil on the ground was once a sea urchin like the modern specimen he is holding. Photo by Dudley Simons.

was aware of their true nature. Others, though, have observed the phallic nature of many of these spines, and related them to a distinctive part of Saint Paul's anatomy.

While the eyes of some of Scilla's Italian and French contemporaries were slowly being opened to the true nature of fossils, Gesner was unable to shake off completely the entrenched medieval views. He still attached some importance to the perceived "powers" of fossils, views entrenched in the human psyche for millennia. These powers were, in part, of a practical nature, with some fossils, including sea urchins, believed to be of direct medicinal value when powdered and ingested. John Woodward in his book *An Attempt Towards a Natural History of the Fossils of England* (published posthumously in 1729), observed that the fine chalk found inside some fossil sea urchins was most efficacious in settling the stomach. He reported that *Echinus marinae* (as he called the fossil urchins) dug out of chalk pits in Kent were often sold by the quarrymen to seamen, who paid well for these "Chalk-eggs." The fine chalk they contained was "one of the finest remedies for subduing acrid humours of the stomach":

> Those who frequent the Sea, and are not apt to vomit at their first setting, fall frequently into Loosenesses, which are sometimes long, troublesome and dangerous. In these, they find *Chalk* so good a Remedy that the experienced Sea-Men will not venture on board without. They chiefly make use of that which is contain'd in the Shells of Echini Marini; which indeed is usually very fine and pure. These are dug up commonly in the *Chalk-Pits* on each side of the River, of Purfleet, Greenhyth and Northfleet, where the Chalk-Cutters drive a great Trade with the Sea-Men, who frequently give good Prices for these Shells, which they call Chalk-Eggs.[18]

Given that the fine chalk powder is calcium carbonate, its effect on neutralizing stomach acidity is not overly surprising. But to most sixteenth-century naturalists the powers possessed by fossils were altogether much stronger and more pervasive. Thus, the sea urchin's five-pointed star was seen as reflecting stellar influences. Many of these naturalists were strongly influenced by the Renaissance Neoplatonic philosophy that saw a rigid hierarchy of the universe, from one God at the highest level, through lower levels that were "emanations" from God. The world was the lowest level of the universe, as it is the farthest from God and is both less real than the rest of the universe and less like God. Powerful forces were seen as emanating from the stars that were scattered in these higher heavenly levels.

Similarly, the star-shaped segments that form the stem of some pentacrinitid fossil crinoids (sea lilies) owed their shape to stellar influences. Gesner called these fossils *Asterias separatus* (separate stars) and classified them as "forms like heavenly bodies (stars)." As such, these objects were thought to possess restorative powers. Thomas Nichols "sometime of Jesus College, Cambridge," writing in 1652 in his *The Lapidary; or, The History of Precious Stones*, commented that "Asterias or Starre-stone is worth two crowns, since it has power to give victory over enemies, is good against Appoplexies, and by the very touch of the body to hinder the generation of worms." In Germany they were called Sternstein or Asteroiten, meaning "star-stones." Their perceived medicinal powers are shown by the use of just four grams of powdered "star-stone," which, when dissolved in water, was thought to protect against not only intestinal worms, but also the plague, as well as helping lung, liver, blood, and genital problems. A star-stone hung in a room was believed to neutralize poisonous motes that floated in the air, which were thought to cause a falling sensation in your sleep.[19]

Yet, notwithstanding the insights of some of the Italian naturalists, and Gesner and Palissy in mainland Europe, ideas in England on the origins of fossils were far less enlightened. More than one hundred years after Gesner's book on fossils was published, Dr. Robert Plot (1640–1696), the first Keeper of the Ashmolean Museum in Oxford, was still not convinced that fossils were the remains of once-living creatures. In his *Natural History of Oxfordshire* (1677), he included some discussion on "Formed Stones," as he called fossils. While he saw some resemblance between some fossils and living organisms, he failed to be convinced of their organic origins. To Plot, these fossils were "*Lapides sui generis*, naturally produced by some extraordinary plastic virtue latent in the Earth or Quarries where they are found."[20] He believed that some fossils, though, came from the stars (see chapter 12).

The same view was expounded by Meric Casaubon (1599–1671) in his book on "spirits, witches, and supernatural operations" published posthumously in 1672, also viewing fossils as imitating living organisms:

> I am a great admirer, I profess it, of a stone, which is not very rare. Many call them *Thunder-stones* [see chapter 9 of the present text]. Naturalists by many pregnant instances, do maintain, that neither Sea, nor Land doth produce any thing, but is imitated and represented in some kind, by some kind of *fossile* in the bowels of the Earth: (whence so many bones of Fishes, yea whole Fishes, imperfect, as to the form, but perfect stone, are found,

and digged up out of the Earth, even upon high hills, far from the Sea: some my self have, and look upon, when occasion offers its self, with pleasure, and admiration) these things considered, I think it is possible, these stones may be nothing else, (but even so, well deserving some kind of admiration) but some kind of *fossiles*; nature aiming by them at the representation of somewhat that doth live, or grow, either in the Sea, or upon the Land. But I forget my self.[21]

It was not until the Royal Society's curator of experiments, Robert Hooke (1635–1703), put his mind to the question of fossils that English naturalists began to accept their organic nature. One of the greatest minds of the seventeenth century, adept not only in astronomy, physics, chemistry, architecture, microscopy, horology, and music, Hooke was also an incredibly perceptive geologist.[22] In his famous book *Micrographia* (1665), he used fossils as one of his examples to demonstrate the great usefulness of the microscope. Examining a thin section of some fossil wood, he remarked that it was fossilized due to its "having lain in some place where it was well soak'd with petrifying water," this having impregnated the wood with "stony and earthy particles." Hooke's views on fossils derived in part from the simple, rational reasoning that any explanation other than that they were once living forms was basically "quite contrary to the infinite prudence of Nature"—it was just unthinkable that they could be anything else.[23]

> All these and most other kinds of stony bodies which are formed thus strangely figured, do owe their formation and figuration, not to any kind of *Plastick virtue* inherent in the earth, but to the shells of certain Shell-fishes, which, either by some Deluge, Inundation, Earthquake, or some such other means, came to be thrown to that place, and there to be fill'd with some kind of Mudd or Clay, or *petrifying* Water, or some other substance, which in tract of time has been settled together and hardned in those shelly moulds into those shaped substances we now find them.[24]

Hooke's experience with fossils was engendered by a youth, like mine, spent wandering the beaches under towering cliffs of crumbling chalk.[25] But whereas my paleontological perambulations were carried out in Sussex, his were some thirty-seven miles (60 km) to the west, on the Isle of Wight. Prominent among the figures of fossils that he drew (and which were posthumously published[26]) were flint fossil urchins that he no doubt found washed up on the beach; fine specimens can still be found that way today. Centuries ahead of his time, Hooke not only recognized that they were the remains of once-living animals, but identified two basic types. One type he called "Helmet-stones," the other "Button-stones." The former are what we now call irregular urchins, the latter regular urchins. Seeing the close similarity between these forms and living urchins, Hooke argued that

> any one that will diligently and impartially examine both the Stones and the Shells, and compare the one with the other, will, I can assure him, find greater reason to perswade him of the Truth of my Position, than any I have yet urged, or can well produce in Words; no Perswasions being more prevalent than those which these dumb Witnesses do insinuate.[27]

When I was collecting fossil sea urchins in the mid-twentieth century, just along the coast from where Robert Hooke had found his three hundred years earlier, there was no doubt that they had once been living, crawling, feeding, reproducing creatures living on and within white chalky mud. These fossils were highly variable in appearance. There were the sort with which people are most familiar today, urchins that are most often seen in rock pools—round, crusty balls covered by a hideous array of spines, whose sole intent in life seems to be to propel themselves into some unsuspecting person's foot. When fossilized, these sorts of urchins are Hooke's "Buttonstones." But such urchins generally make poor fossils. After death (more often than not caused by some predator, such as a fish, snail, or starfish) the remaining pieces of shell often disintegrate before they can be fossilized. But there is another sort of sea urchin, much more secretive, that lives its entire life hidden from view.

The floor of the chalk seas was a soft, white ooze into which these furtive creatures, the so-called irregular sea urchins, burrowed. Rather than being covered by just a few large spines, these forms were quite different. Often egg-shaped, and with a shell equally as fragile, these urchins were covered by hundreds of tiny spines, making the animals look for all the world like tiny, anemic, aquatic hedgehogs. After death the spines usually fell off quite quickly, and so they have been rarely found on the fossil. Like all urchins, the main characteristic of the shell is the highly distinctive set of five grooves called ambulacra that radiate outward from the center of the shell, forming a distinctive star-shaped pattern. In life these were crucial to the urchins, as they carried double rows of closely spaced pores that were pierced by fleshy tentacles called tube feet. These tube feet served a whole range of functions. They were used for respiration, for sensory purposes, or

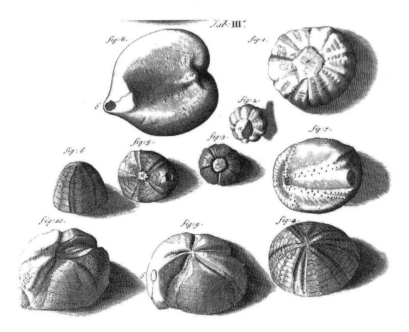

Robert Hooke's drawings of urchins, known by him as "Button-stones" and "Helmet-stones," based on fossils he found as a boy on the beaches of the Isle of Wight. These and other illustrations were drawn by Hooke to accompany a series of lectures, A Discourse of Earthquakes, that he gave to the Royal Society in 1668. They were published posthumously by his friend Richard Waller in 1705.

for secreting a thin film of mucus that formed a tent around the urchin in its muddy burrow. Among the animal kingdom it is only the echinoderms (urchins, starfish, and the like) that have evolved this five-fold, or pentamerous, symmetry.

Searching for fossils of these urchins has meant looking for different things. They took on different shapes. Some were high cones, their helmet shape inducing Hooke to call them "Helmet-stones." The scientific name of the most common form of these "Helmet-stones" is *Echinocorys*. Others, called *Micraster*, possessed an almost perfect heart shape, hence their common name of heart urchin. Sometimes fossil sea urchins are preserved as a calcitic shell within the chalk. More often than not they are found as flint molds of the hollow inside of the urchin's shell. To any unashamed collector and hoarder of rocks, what could be more prized than a perfectly formed stone, shaped like a heart, with a five-rayed star incised on its surface?

As I became progressively more unsuccessful in my quest for these elusive treasures, I changed my tack. Why dig through a quarry looking for a few fossil flints, when at the foot of the cliffs into which the sea gnawed lay almost as many flint pebbles as stars in the universe? Why not take advantage of the help given by millions of years of erosion of the chalk cliffs that lined the English Channel, as Hooke had done centuries earlier? These formed the pebbly beaches of Brighton, cursed by generations who have had to endure wading through this mass of oversized flint marbles ever since the English took to leaping into the frozen sea with such delight. Yet to me it seemed that the beaches might, perhaps, be the source of my ultimate fossilized prize: an urchin.

I can't begin to think about how many pebbles I must have looked at in the desperate hope of finding an urchin. Although you may think that if you've seen one pebble from Brighton Beach you've seen them all, let me tell you that the variety of shapes, sizes, and colors was as perplexing as it was sometimes rewarding. A few pebbles had channels and tiny "caves" within them, in which minute, sparkling crystals of quartz had grown. Others had holes in them; so because they were unusual, they were inevitably collected. And very occasionally, often after many hours of searching, I would be rewarded by finding perhaps a badly weathered and worn helmet-shaped *Echinocorys*, or sometimes a pea-sized urchin called *Offaster*. But I don't ever remember finding in these pebbles the most sought-after fossil urchin of all: the beautiful, heart-shaped *Micraster*.

Such beach collecting was fraught with many surprising encounters, not the least of which were with naked, usually portly Sun worshippers. They lay like pink, bloated elephant seals on the nudist's portion of the beach on which I searched, worshipping their solar deity in all their glory. Convincing some of the more mighty ladies that I was not an urchin myself, but was in fact looking for urchins—"no, not the two-legged kind, madam"— was sometimes less than easy. After all, collecting fossil sea urchins was not the sort of thing that one boasted about in the working-class parts of Brighton where I lived. I mean to say, who in their right mind would collect something as obscure as a fossil sea urchin? So collecting fossils became something of a clandestine operation. That meant I wouldn't have to explain to people what I was doing, or why I wanted to fill my bedroom with heart-shaped stones and little domes of flint.

Things haven't changed much, forty years on. I still collect fossil sea urchins. Trying to convince somebody that, apart from their intrinsic attractiveness, they are actually useful in evolutionary theory and paleobiogeographical studies becomes something of a wasted exercise. I find that one of the quickest ways to kill conversation at parties is to tell people what I do for living: "I study fossil sea urchins, and today I've been measuring the size of the anus of a particularly fascinating specimen . . ." By this time I'm usually talking to myself. *Why doesn't he do something useful for a living,* they are probably thinking, *like selling real estate, or fish and chips?* Why not indeed?

Oceans of flints were searched for fossils through languorous summer days, fields of flint scoured in dank days of autumn; but in frozen winters flint sometimes ruled my life in other, less pleasant ways. I still carry the scar on my forehead where I tried to head-butt a broken piece of flint, after being catapulted out of a sledge pulled too quickly along an icy, flint-bedded track. This track that my forehead knew so intimately ran along the outside of a flint wall bounding the territory that my friends and I prowled around when it was too cold to search for fossils—when the Earth was covered in snow, or the soil too entangled with ice to yield up its frozen stars. To us, that wall was a barrier of such absolute power and intensity that only the brave or foolhardy would ever scale its knobbly heights and attempt to cross it. On our side lay the comforting security of homes and gardens—on the other a tangled wilderness within which lurked the terrors of our imaginations; for it was not just the mere physical presence of the wall that controlled us, but the unspoken cold fear of what lay hidden on the other side: an isolation hospital for infectious diseases.

To our young eyes it rivaled the Great Wall of China and made Hadrian's Wall seem like a flimsy fence thrown up in haste on a Saturday afternoon before new, and rather unpleasant, neighbors turned up. For our wall was an eructation of flint, spawned of the very substance of the hills—the eternal debris that would withstand eons of weathering and erosion, and all the forces that nature could throw at it. This towering, massive structure (it was, after all, probably six feet [1.8 m] high) was not only almost unscalable—if, horror of horrors, it had to be climbed to retrieve a soccer ball which had foolishly been kicked over it. Worse than that, the wall was topped by jagged rows of flints. Who needs broken bottles when these cherty slivers, with edges keener and more deadly than any razor, were in such plentiful supply? Given that this wall must have run for over a mile, it is sobering to think of the poor souls whose job it must have been to break the flint nodules and turn them into weapons aimed at maiming the unwary trespasser. But all they were doing was emulating a tradition that their ancestors had been carrying out in this area for tens, even hundreds of thousands of years—using flints for a multiplicity of uses: for cutting, as arrowheads, scrapers, and hammers. Before humans learned how to smelt metal, these flints were one of the few materials that could be fashioned into tools, along with wood and bone—and they were far and away the best. A large amount of each day must have been spent by these early people, scouring the ground for the best flints. Here, though, their collecting had a utilitarian motive.

In more recent times flint has been used to build not only walls, but also houses, churches, and barns in parts of the country where it has accumulated over millions of years. Because the chalk in which it forms is far softer and is also much more readily subject to chemical erosion, over the millions of years that the chalk downland has been eroding, flint has become concentrated in the soil. Known locally and creatively as clay-with-flints, this deposit accumulates on the hills, then works its way into topographic depressions and valleys.

The silica gel from which the flint was formed, and which often seeped into a dead, empty urchin shell, owes its existence not only to the presence of fields of sponges, but, a little surprisingly, to forces operating from outside the confines of the earth. The chalky mud that makes up most of the sediment laid down in these ancient seas tells us of a time when the Earth was much warmer than today, and one that was a true "greenhouse world." Carbon dioxide (CO_2) levels were about four times higher than at present (without the help of humans), and the oceans acted as enormous sinks that soaked up this planet-warming gas. The microscopic marine plants that live in oceans (known as phytoplankton) were living a life of unadulterated hedonistic luxury during this time, proliferating in this warm CO_2-rich world and leaving their remains as chalky ooze on the seafloor after they died.

But the Earth's climate is never at rest. We know from the ravages of El Niño how rapidly and destructively the Earth's climate can change over periods of just a year or two. It has been like this for a long time. If you don't believe me, take a ferry to the Isle of Wight in southern England. Go and stand on the beach on the southwestern end of the island, a little to the east of The Needles, where the young Robert Hooke spent much time looking for "Button-stones" and "Helmet-stones." With your back to the sea gaze up at the great towering chalk cliffs that surge into the sky, and you will see evidence right before your eyes that the Earth's climate was just as fickle 80 million years ago as it is today. For the cliff looks like one gigantic peppermint humbug, with regularly repeated stripes of white and black. The latter represent thin beds of flint that can be traced, in some instances, for hundreds of miles. These rhythmically alternating layers of soft, white chalk and hard, black flint are the product of rhythmical oscillations in oceanic temperatures, brought about by changes that originate in the vagaries of the Earth's rotation. Such cyclicity in sediments is quite common.

A number of different cycles, called Milankovitch cycles, representing fixed periods of time have been identified during the last few years. Some cycles represent about 20,000 years, others 40,000 and about 100,000 years. The flint stripes in the chalk are thought to reflect the 20,000-year cycles caused by the Earth's slow wobble as it spins on its axis (the precession of the equinoxes). It's a bit like the way a spinning top wobbles as it begins to slow down. The 40,000-year cycle is thought to be due to the cyclical change in the tilt of the Earth's axis, which varies between about 21.5 and 24.5 degrees, causing cyclical variation in the amount of solar radiation reaching the Earth. The 100,000-year cycle is due to variations in the orbit of the Earth-Moon system around the Sun. Short-term cycles that we all experience are the roughly decade-long sunspot cycles. Another, of course, is the annual cycle of the seasons, due to the tilt of the Earth's axis.

All these extraterrestrial effects result in changes to the amount of solar radiation that reaches the Earth. When there's more, the atmosphere and the oceans are warmer. Conversely when there's less, cooling takes place. And it is when these cooler periods occur that the chemical environment is more suited to the precipitation of silica gel on the seafloor which eventually hardens into flint, sometimes inside empty urchin shells. So, rather amazingly, whether the urchin you collect is preserved in calcite or in flint is due to wobbles in the Earth's axis 80 million years ago.

Judging from the large numbers of fossil urchins preserved in flint that have been found by humans, these variations in climate, with many cooler episodes, must have been a common occurrence in the ancient Cretaceous seas, 80 million to 65 million years ago. One reason for the frequency with which these urchin fossils have been found is probably their mode of life. Living completely or partially buried in mud, they were well suited to ultimately being transformed into fossils after death. Whether your mortal remains ever attain this state of almost infinite existence—depending on the vagaries of later weathering and the avidity with which people with hammers take to fossil-bearing rocks—not only is a mixture of luck and good fortune, but also depends on where you choose to die. A butterfly, for instance, that gives up its last metaphorical gasp on a windswept hill in the rain, surrounded by hordes of hungry predators licking their lips and ready to turn the poor insect into their plat du jour, is much less likely to achieve paleontological immortality than an urchin that has lived its entire life entombed in soft, quiet sediments.

Urchins are not exactly the most fleet-spined of animals in life; they propel themselves slowly forward in the sediment by hundreds of tiny spines. Virtually the only outward difference in death, compared with life, is that the spines stop moving. So long as nothing else digs deep into the sediment and disturbs the urchin in its final resting place, there is a good chance that it will eventually become fossilized. If it shuffles off its mortal coil in one of the colder phases, when solar radiation is less than normal, it may well be slowly filled by creeping silica gel. If that is the case, its chances of becoming fossilized will be much improved. Once the gel has hardened into tough, resistant flint, the likelihood of its ending up on the mantelpiece of a most strange, bipedal mammal 80 million years into the future will have been greatly enhanced.

I lived my childhood with a strange, secret passion for these fossil sea urchins. On more than one occasion, as the rain beat down on me in yet another sodden chalk quarry, I asked myself what on Earth I was doing there. Why wasn't I at home sitting in front of a nice warm fire tucking into plates of toast and Marmite and watching *The Lone Ranger* on TV? It's a shame that all those years ago I never put two and two together. I was merely being driven by a genetic imperative that had been imprinted on my species ever since it first drew breath: to hunt for that elusive prey, and then capture it. In other words, to collect, acquire, or otherwise hoard a functionally quite useless item. But unbeknownst to me, I was part of a rich and, as I was subsequently to discover, extremely long heritage of urchin collectors. Had I realized this at the time, then my longing to collect fossil sea urchins might not have seemed quite so strange. I would have come to understand that I was a very long way from being alone in possessing this peculiar trait, given the many thousands of years that humans had been collecting fossil sea urchins. Although their interest in these enigmatic stone urchins was probably far more complex than mine had ever been, like me, what had most probably attracted them in the first place was a five-pointed star.

> Good frend, for Iesus sake forbeare,
> To digg the dust encloased here:
> Blest be yᵉ man yᵗ spares thes stones,
> And curst be he yᵗ moves my bones.
>
> **WILLIAM SHAKESPEARE,** *his epitaph*

 # 4

Skeletons

You are walking along the edge of a plowed field on the chalk downland. A hedgerow of hawthorns, sloe, and beech forms a barrier on one side of you; a muddy field studded with thousands of ankle-wrenching flints of all shapes and sizes slopes away on the other side. Dark gray flint knucklebones with thin white rims like a ghostly skin reflect the afternoon Sun. Suddenly, in this chaotic mass of irregular shapes, something with more regular form catches your eye. The symmetry, curiously enough, is almost jarring. You stop. With your walking stick freshly crafted from a fallen branch of a beech tree, you poke this rounded flint. Not only does its smooth, round form evoke a unique symmetry in this sea of siliceous chaos, but so, too, do the five pairs of delicate lines radiating like rays from a figurative star.

Your curiosity gets the better of you, so naturally you bend down and pick it up. You brush off the loose clay that smears half of it. It nestles neatly in your hand. But what is it? Maybe you recognize it right away. You may well be versed in the occurrence of *Micraster cortestudinarium* in the Upper Chalk, and know instantly that you are holding a fossil sea urchin. Then again, maybe not. More likely you think that you have a pebble with the fossilized imprint of a starfish splayed across its surface. Either way, because you live here and now, perched on the razor's edge of the twentieth and twenty-first centuries, there's a good chance you realize that this was once some kind of living marine creature, now somehow turned to stone.

You try and remember where you have seen one of these before. You think skeleton. Why skeleton? Then it all comes back to you. Skeletons of two young women that you had seen once in the Brighton Museum. Someone else had picked one of these up a few thousand years before you did. For some unknown reason each woman had been buried with a fossil sea urchin and little else. Fossil collecting clearly has a rich heritage. You wonder what these women would have made of such a strange object. Did they have a name for it? You pop the urchin in your pocket, exhibiting the self-same impulse to collect the strange and the unusual that they did.

On those dark days when winter hung heavy and it really was too cold and wet to venture out hunting for urchins, I was sometimes drawn in my youth to other places to assuage my collecting lust. It is late afternoon on a cloudy, windswept day and nearly dark. The wet streets seemingly try hard to glisten on such a drab afternoon. The rain starts to come down harder, providing even more incentive to walk faster. Just a few hundred yards to go. A car speeds past, shooting out a muddy spray of water. Will all this effort be worthwhile? What else to do, though, on such an evil day but visit that temple of officially sanctioned collecting—the local museum. At this time of the year it is one of the few places offering free entertainment.

Finally, into the building and out of the nagging rain. The difference is stunning. The roar and hiss of passing vehicles and the noise and clatter of the town suddenly disappear. Here all is dry; all is dust. The only sound comes from the water dripping off sodden clothes onto the cold floor, and the buzzing of lamps. But it is not the dull yellow light that stays in the memory. It is the smell. That very special, but quite indescribable smell of English municipal museums in the mid-twentieth century. But once smelled, always remembered. Not a smell to want to bottle and cherish for its own sake, but a smell that is redolent of many generations' urge to collect.

The building suffused with this smell was the Brighton Museum, housed, arguably, in one of the strangest of all buildings in England. In 1786 the then Prince of Wales, later to become King George IV, decided to establish a base outside London. He chose an unpretentious house situated in a flat valley that runs through the center of a tiny fishing village, called at that time Brighthelmstone. In debt to the tune of a quarter of a million pounds (due, in his opinion, to an ungrateful father and an ungenerous Parliament), the prince was forced to live in relative modesty. However, it only took him a year before he enlisted the services of Henry Holland, who was to design an enlarged, grander house, to be known as the Marine Pavilion. It took a mere three months to transform this original "two up, two down" into a pair of villas of Palladian splendor, linked by a striking domed rotunda, partly encircled by six Ionic columns, and surmounted by classical statues.

Over a number of years the interior was transformed into a palace with an opulent Chinese character. But it was not until 1817 that John Nash created the Indian character of the exterior of the Royal Pavilion, as it became known, with huge domes looking for all the world like gigantic, elegant onions. Indirectly this activity played a role in the discovery of the first dinosaur remains in Sussex. The quarry that was engaged to supply the stone for the rebuilding of the Royal Pavilion, near Cuckfield in Sussex, became greatly enlarged as a consequence. It was during this excavation that the teeth of the dinosaur *Iguanodon* were discovered by Gideon Mantell in 1821.[1]

The prince's horses were housed in stables constructed a little earlier (1805) at the staggering cost of £55,000 and which formed part of the complex of this amazing building.[2] This equine Ritz would have been the envy of virtually the entire human population of the country at the time. There cannot be too many museums whose precious objects grace rooms where horses once contributed to the fertilizer trade, for the Royal Stables and Riding School are now occupied, in part, by the museum.

While in more recent times the museum may be better known for its sofa designed by Salvador Dali in the shape of Mae West's lips, it had, through much of the last century, a fine display of much older archaeological objects that had been uncovered from countless sites in the region. Indeed, its archaeological collection is among the best in municipal museums in the country. The two objects that have forever stuck in my memory are a beautiful amber cup and some skeletons. These objects seemed, at the time, to be as far removed from me as the fossils I collected. The amber cup had been found one hundred years before I first set my eyes on it, at Hove, in the only large barrow (a burial mound) to have been found on the coastal plain. For as long as anyone could remember, until it was excavated in the mid-nineteenth century, hundreds of young children would dance around this 220-foot-wide (66-m), 12-foot-high (3.6-m) hill every Good Friday playing "kiss-in-the-ring."[3] This is a skipping game that was mainly performed by children on Good Friday. It was almost invariably carried out in close proximity to a barrow. Even today, such skipping is carried out on Good Friday in Alciston, a village having the densest concentration of barrows in Sussex. It has been suggested that such games came to England with people who introduced the concept of burials in barrows, in other words during the Neolithic. "Communal skipping on or near barrows . . . does seem to be connected with barrows in the minds of the people who do it . . . I think we may consider present-day skipping as a far-off descendant of the sports and games played at burials and, because continuity of tradition is almost ineradicable when people wish it to be so, possibly at barrow funerals."[4] Moreover, it has been suggested that such skipping carried with it a renewal-of-life motive and was originally performed during burial rituals.

When the Hove barrow was destroyed in 1856 in order to uncover its hidden secrets, it was found to contain an oak coffin shaped like a dugout boat and cut from a single tree trunk. Little was left of its original inhabitant, but with this long-departed soul had been buried a magnificent stone axe (one of the finest found in Britain), a bronze dagger, a small whetstone, and the amber cup.[5] This cup, the size of a large coffee mug and complete with handle, had been carefully carved from a single piece of red amber and surely must have been a prized possession of the person buried in the coffin. Similar grave goods found in Scandinavia have been linked to Thor, the Norse god of thunder, and his mighty hammer Mjöllnir. As I stared at this glowing amber cup I had (not surprisingly) no idea that my quest for fossil urchins and these four objects found in a rotting oak casket were to be related.

Of much more interest to me at the time were the human skeletons. These, I am sure, were the first skeletons I had ever laid eyes on. So it is no surprise that they became deeply etched into my psyche. Thousands of years old, they had been excavated from two shallow graves that had lain slap-bang in the middle of the Brighton Racecourse.[6] To find skeletons on display in a museum like this was not really that unusual, given the rich archaeological heritage of the district. And after all, any museum worth its salt must have its requisite human skeletons on display: dry, brittle bones exhibited to excite, enthrall, and entrance the visitor. But these remains were different—not the skeletons themselves, but the meager grave goods found with them.

When I first peered closely into the display cases in which these bodies now reposed (cases looking a little too much like transparent coffins, for my liking), I was transfixed. It turned out that I wasn't the first person in these parts to have been seduced by the lure of fossil sea urchins, for one had been found with each body. Had they, like me, been fossil collectors? Or had another person found the fossils and at some later date, for some strange reason best known to that person, placed them in the graves? Either way I was enthralled by this inexorable link with the past—a thin thread of contact with people who had lived thousands of years earlier. Suddenly these dry bones came to life. Some people in the prehistoric past had shared my passion. And they were decidedly more successful than I had been at finding fossil urchins. As I later discovered what these people might had been doing in this area so long ago, I realized how my own activities were not so very different from these ancient fossil collectors, for they, too, had wandered the same hills over which I so often tramped.

Early on Sunday mornings, when the long days were already heavy with the droning of bees, my friends and I would congregate by the same flint wall that separated us from the other world of our fears: a world where people forever lay dying of hideous, incurable diseases. Armed with a battle-scarred bat and a ball that had long ago been beaten into submission, we would set off, trekking up the curiously named Bear Road. We couldn't escape that wall, though. It never left us. It pursued us all the way. Closer to the top it became even more frightening. We knew that before we reached the safety of the hilltop it hid even worse terrors. For behind its crusty flint exterior lay a cemetery. While skeletons in a case in a museum bothered us not one jot, the thought of the serried ranks of graves lurking behind the wall sent a chill down our young spines. When we finally escaped the clutches of the wall and reached the top of the hill, we breathed a sigh of relief—we had reached the racecourse.

In terms of its location, the Brighton Racecourse is probably one of the most spectacular in the entire country. Shaped most appropriately like an open horseshoe, the course curves along a ridge that is flanked by short valleys (called coombes) running south to the sea, and north to the scarp of the South Downs. These coombes are the product of periods when abundant flowing water gouged deep, rounded valleys in the soft chalk. The water most probably derived from permafrost that had smothered this part of the country some twelve thousand years before, during the climax of the last great Ice Age. As it melted it spewed water south to the sea and north to the river valley. Left behind after all this aqueous excitement was the narrow, curving ridge along which horses now gallop.

As they reach the four-furlong mark and come onto the home straight, the horses head south. After passing the finish line, they break into a canter through what, 4,500–5,500 years ago, was a Neolithic causewayed camp known today as Whitehawk Camp. This is where the Brighton Museum skeletons were found; there is now little evidence of the four concentric oval rings of ditches that formed this site. Like our ancient ancestors, who may well have congregated here for worship, so my friends and I gathered at the top of this hill to conduct our own religion. No, not horse racing—cricket.

These games were played on the soft, warm days of summer that seemed to spread out into never-ending twilight. It was certainly not a place to be in the frozen pit of winter, with vicious winds snarling across the exposed hilltop. Who in their right mind would be up there then? But thirty years before our cricket matches, one man had been. A man, actually, of very sound mind who numerous times had walked briskly down the racecourse on cold winters' mornings—a man who was to discover, beneath the frozen turf, some bodies.

Cecil Curwen plunged his numbed hands as deep into his pockets as he possibly could. Even with a thick overcoat, hat, and scarf he still felt frozen to the marrow. The footprints that he left as he walked across the frosted turf were less intrusive than the hoofprints that would pockmark the turf in a couple of months' time. Curwen wasn't quite sure if the last few weeks he had spent up here had really been worth all the effort. His hopes for a restful festive season had been dashed early in the December of 1932 when a letter had dropped unceremoniously onto his front-door mat. The letter had been sent by the Museum Subcommittee of Brighton Corporation. This subcommittee was, in essence, just one man—Herbert Toms, curator in charge of the archaeological collections, and well known to Curwen. Some twenty-five years earlier, when Curwen was just eleven years old, he had been elected at Toms's urging to the newly formed Brighton and Hove Archaeological Club.

The letter informed him that the Lessees of the Brighton Racecourse had decided, in their infinite wisdom, that at the beginning of the 1933 racing season the horses should have an extended pulling-up area. The only problem was that the proposed extension to the racecourse ran straight through the two outer ditches of Whitehawk Camp, one of the more significant of such Neolithic sites in southern England. The causewayed camp is roughly circular in outline and has a diameter of about 333 yards (300 m). It consists of four concentric ditches interrupted by causeways.

Found high on chalk downland, causewayed camps were first constructed about 5,500 years ago. As such they represent some of the earliest modifications of the landscape undertaken by ancient Britons. These structures consist of a sequence of incomplete concentric ditches broken by sections of undisturbed chalk. Similar structures are also found in Germany and Belgium, suggesting that some aspects of the culture associated with these Neolithic camps in England emanated from continental Europe.

Exactly what these causewayed camps represent has been hotly debated. Undoubtedly they were of great significance to Neolithic peoples; sixteen have been recorded from southern England alone.[7] Some archaeologists think that they might have been fortified villages or sites for periodic fairs or ritual feasts. Others have suggested that they may have been monuments involved with the establishment of power and the differentiation of developing Neolithic societies through the control of labor, production, and ritual. But Alasdair Whittle, writing on Neolithic societies in Europe, believes that such camps "were the focus for intensive, participatory ceremonialism, which reworked ideas about the integration of separate communities, and brought the past into the present."[8] In a period when the population was probably widely dispersed in small communities, there is little doubt that these enclosures represent communal meeting places of some kind. Curwen thought that they were the headquarters of pastoral people, and that the numerous causeways facilitated driving their flocks and herds into the central enclosure.

Fortunately the Whitehawk causewayed camp was listed under the Ancient Monuments Act, so archaeological excavations had to be carried out before the racecourse could be extended. As the author of *Prehistoric Sussex*, published just two years earlier, Curwen was the obvious choice to undertake the work. Born in Peking, China, in 1895, he had been educated at Rugby and at Gonville & Caius College, Cambridge, before training to become a physician at St. Thomas's Hospital, London, shortly after the First World War. While his medical work paid the bills, Curwen's real love in life was archaeology, a passion he undoubtedly inherited from his father, the renowned archaeologist Eliot Curwen.

Together father and son undertook most of the major archaeological digs in Sussex, particularly the South Downs, in the years between the two world wars. They excavated a large flint mine on Harrow Hill in 1924–25 and the Caburn hill fort in 1925–26, as well as the Neolithic causewayed camp and Iron Age hill fort at The Trundle, near Goodwood, in 1928–29, and the Roman site at Thunderbarrow in 1932. But it was for more than this that Curwen was chosen to lead the Whitehawk Camp excavation. He was

E. Cecil Curwen, excavator of the two Neolithic burials from the Whitehawk causewayed camp, Brighton, Sussex. Photo courtesy of the Sussex Achaeological Society.

certainly no Indiana Jones. On the contrary, he was renowned for his very methodical methods, recording information with great clarity of thought. It was this attention to detail that was to prove crucial at Whitehawk. As his obituary in late 1967 recorded, "Dr Curwen was a shining example of the great amateur archaeologists who set the standard for the work of the moderns."[9] Toms had little doubt that Curwen was the man for the job.

Curwen probably was delighted as he read the letter. This project would give him the opportunity to continue excavations of the camp that he had started four years earlier. However, the excavation had to be completed by January 21.[10] That gave him five weeks, six at the most. One question was, how to do the work in such a short period? He thought back to those who had worked with him before: Hamilton to measure and record again; Burstow, Holleyman, and James Stuart to help with the digging again.[11] Whether he could persuade them to join him was another question.

As to how he would pay them, he saw from Toms's letter that the princely sum of £125 had been offered for the completion of the work.[12] Not a fortune, but it was sufficient. They would just have to be convinced that they would rather be digging up the Brighton Racecourse at Christmas than enjoying a port and lemon in front of a blazing coal fire. Somehow, though, he persuaded them all, and apart from a short break at Christmas they worked through December into January.

Curwen was possibly a little disappointed by what they found in the first few weeks. They had dug out most of the two ditches that he had planned, one for about 100 feet (30 m), the other 140 feet (42 m), and both about 20 feet (6 m) wide:[13] no mean feat given the frozen condition of the ground. Curwen usually turned up at the site later than the other men. But he had a good excuse—he had spent many nights photographing the ditches. This was not so strange as it seems. Given that even when the Sun shone at this time of the year it was very low in the sky, and the light levels pitifully low, Curwen reasoned that it might be better to take long-exposure shots by moonlight, "just at [the] time the full moon was sailing high in a clear and frosty sky."[14] His reasoning was spot on. The stark contrast between the ground illuminated by the white moonlight meant that he produced excellent photographs.

Curwen left no account of the actual discoveries of the skeletons. We can only imagine what the scene might have been like when he arrived one day to find more activity at the site than usual. Hamilton, perhaps, stands on the edge of the ditch, looking down. As he gets closer Curwen may hear two or three voices talking at once. Usually there was little conversation when the men were actively digging—just a few grunts. Perhaps Hamilton turns and catches sight of Curwen coming toward him. He waves in a frenetic, beckoning manner. Curwen doubles his pace. As soon as Hamilton thinks Curwen will be able to hear him, he tells him that they've found some bones. But this time they're not deer bones.

The one interesting find that the party had made so far was a dismembered corpse of a roe deer. It was littered with snail shells that had been carefully placed on the back of the animal, which lay in the bottom of a small hole in the ground.[15] Curwen was sure that the hole had been dug especially to receive the carcass. Such a careful burial, in his opinion, implied a ritual killing and burial—perhaps some sort of foundation sacrifice, if the hole was a posthole to support a building.

Peering into the ditch with the excited Hamilton, Curwen would have been presented with the sight of the remains of a skeleton that was undoubtedly human. It had been buried in the middle of the ditch, in a semi-prone position on its left side, with its legs tucked tightly under its body. All that had been found in the way of grave goods, probably a little to Curwen's surprise, was a single fossil sea urchin: the helmet-shaped *Echinocorys*.[16] Later examination revealed that the skeleton represented the remains of a small young woman aged between twenty-five and thirty years.

This was not the only surprise that these ditches had in store for Curwen and his crew. As they were coming to the end of the excavating, they were working a little further east in the same ditch. Digging easily through the loose chalk rubble, the men suddenly struck a number of large chalk boulders nearly one foot across. As they continued digging they began to find pieces of bone: ribs, arm bones, and then a skull. It was another burial. But more care had been taken with this one, for they soon realized that this body had been buried in a more defined grave.[17]

The skeleton lay within an elongated oval that had been surrounded by ten large and a few smaller blocks of chalk. On top of the soil that had covered the body was a layer of charcoal; subsequent radiocarbon dating indicates the burials took place about 4,700 years ago.[18] Not all the bones lying within the circle of chalk blocks were large. As Curwen and his helpers continued their excavation, they began to uncover paper-thin bones. It was a wonder that they had survived. As the men cleared the chalk rubble from around the bones with extreme care, it became apparent to them that more than one person had lain in the grave. Buried with the adult (who later turned out to have been a female aged between twenty and twenty-five years) was a baby. Its diaphanous fragments of bone lay close to the woman's abdomen, suggesting that it may have been a late-stage fetus.

If Curwen and his helpers thought that this burial site would yield a bonanza of grave goods, they were very much mistaken. All that accompanied the skeletons of the woman and her infant were two simple pieces of chalk into which holes had been bored, perhaps to allow her to thread twine through them to wear as a necklace; the lower half of the radius of an ox; and two fossil sea urchins, again both *Echinocorys*.[19] One fossil, perhaps, for each body. Many archaeologists would have dismissed the presence of the

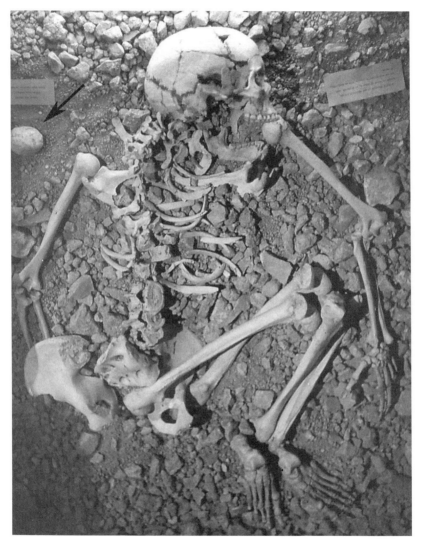

Neolithic skeleton excavated by Cecil Curwen from Whitehawk Camp, with her grave good of a lone fossil sea urchin (*arrow, upper left*). Photographed when on display at the Brighton Museum. Author photo.

fossil urchins as coincidence. After all, the bodies had been buried in chalk, from which the fossil came. But not Curwen. He was convinced that the fossils were placed in the graves deliberately. He did not record finding any other urchins during the period of excavation, other than the three found with the skeletal remains.

Two questions are posed by these graves: first, of what possible significance to these Neolithic people could something as strange as a fossil sea urchin be; and second, how did these women, one with a child barely a few weeks old, meet their deaths? For most societies who bury their dead it is standard practice to send them on their way with grave goods of all shapes and descriptions. Personal adornments, swords, food, certainly—but a fossil? Perhaps the fossils had been put there, as grave goods usually were, to accompany the souls on their journey to the afterlife. But what possible reason could these people have had for taking these fossils along with them on their long journey? Were they simply as fascinated with fossils as I was? Were the fossils mere curios to which these women were especially attached in life? If this was so, it might lend further support to the view that humankind's propensity for collecting strange and interesting objects has a rich and prolonged history. So maybe I wasn't so strange after all. Or perhaps the urchins also held a much deeper significance.

Apart from placing a few blocks of chalk around the female and child, these burials showed little sign of having been made with much reverence. Curwen was appalled by how these poor souls were buried. Within the ditch in which the bodies were found, other evidence that shocked him came to light.

"Life at Whitehawk Camp," he later wrote in his book *The Archaeology of Sussex*,

> must have been at a very low level. The excavations . . . disclosed some sordid secrets. In the middle of a stretch of this ditch 50 feet long between two causeways, some one had squatted for a sufficient length of time to leave a good deal of litter. He had made a fire at which he had cooked his food, for beside a large pile of ashes we found the fragments of several pottery vessels, some animal bones and nearly 100 burnt flints or "cooking stones."[20]

But Curwen found some much more disturbing finds in with this debris.

> Among the animal bones we found parts of the brain-pans of two human skulls, three small fragments of which had been charred by the fire, while scattered about the rest of this part of the ditch we came across three more human brain-pans and one or two other bits of human bone. All the five individuals here represented were young; the oldest was not much, if at all, over twenty; the youngest was six.

He could only come to one conclusion—the person (or persons) who had been cooking in this ditch must have been cannibals.

> Not content, however, with living amid this filthy litter, he must needs bury his young wife and infant at the end of his little ditch, within 20 feet of his hearth. There we found her lying half on her face, with her knees drawn up, and her new-born child beside her in a kind of grave enclosed by large blocks of chalk.[21]

The two fossil sea urchins were buried with them, "perhaps to act as charms," Curwen reasoned. But charms against what? Against perceived evil spirits, perhaps?

The manner in which the other skeleton was buried disgusted Curwen even more: "It looked as if she had been flung into the ditch with the other refuse, for she was lying there half on her face with one arm thrown out behind her and her knees doubled up, with no sort of prepared grave."[22] Her only token of care or esteem was the fossil urchin.

When Curwen returned to undertake further excavations in 1935, he made another discovery suggesting that some of the deaths at this site might have been sacrificial. A deep hole was uncovered on the edge of a causeway of the third ditch; it might have been intended to hold a tall pole. In the upper part of the filling lay the skeleton of a small child, about seven years old. With him were a few shards of pots and a piece of chalk bearing incised lines.[23] Curwen wondered whether the child had been sacrificed, and then buried at the foot of a wooden pillar.

If Curwen was correct, then could the other Whitehawk burials have been part of similar rituals in which people were ceremonially killed and placed into a rough cist of chalk blocks? Establishing the cause of their deaths has not been attempted, and given the nature of the material used for the burial chamber, it is unlikely that any definite conclusions could be reached. Certainly, Whitehawk Camp was not primarily a burial site, but an important meeting place. Why were just these few bodies buried here, unless they did indeed play their own, sacrificial part in a ritual? Possibly the placement of fossils with the bodies occurred because the objects themselves were of particular significance to the dead people. However, it would seem unlikely for such an altruistic action to have been carried out by those who had killed the women and child, if indeed these people had been sacrificially killed—unless, that is, a particularly potent symbolism was attached to the fossils.

There is, though, other evidence in support of ritualistic killing in Neolithic times in southern England. Windmill Hill, another causewayed camp, near Avebury in Wiltshire, has yielded human remains found in ditches.[24] Like Whitehawk, these were just skulls of children and infants, but spaced at regular intervals. Whether they all died of natural causes or were human sacrifices is not clear. However, at other Neolithic sites human remains have been found that had been bound tightly before death. It has been suggested that at Windmill Hill, and maybe also at Whitehawk, the children were buried in the earthworks as a foundation sacrifice.[25] Presumably this was because, for some reason best known to these people, they believed such acts bestowed good fortune on the site. Human sacrifices found in bogs in Denmark may have been associated with cults of regeneration and fertility.[26] The widespread habit during the Neolithic of burying the dead in close association with the activities of the living may also have reflected beliefs in the supernatural presence or powers of the ancestors.[27]

But Curwen was not the only person finding bodies buried with fossil urchins on the South Downs at this time. So, too, was John Henry Pull (1899–1960). A postman by occupation, Pull actively excavated many archaeological sites there in the 1920s and 1930s. At the time he was regarded as one of the foremost amateur archaeologists in Britain, leading excavations at some of the classic sites on the South Downs, including the Neolithic flint mines at Blackpatch and Cissbury. Unlike Curwen and the other amateurs of the time, though, Pull's background was working-class. Indeed, he is regarded as one of the first serious working-class archaeologists in the country; his predecessors all had been affluent professional men.[28] Yet despite the high quality of his research, Pull was not a member of the archaeological "establishment," and most of his contemporaries were contemptuous of his work, with some even actively discouraging its publication. And one of his nemeses was none other than Cecil Curwen.

Curwen was not the first to discover fossil urchins with Neolithic burials, because during the excavation of a ring barrow at the Blackpatch flint mines in 1928, Pull had uncovered single specimens of fossil urchins from two separate late Neolithic Beaker burials.[29] Pull's attempts to get his reports on his early excavations at Blackpatch published by the Worthing Archaeological Society, of which both he and Curwen were members, were actively thwarted by some of the society members.[30] Among these was Curwen, a member the society's Earthwork Sub-Committee. Realizing the significance of Pull's excavations, the members of the Sub-Committee did want the report published—but they didn't want Pull's name attached to it.

The inference that many drew from this was that it had nothing to do with the quality of the report, but was simply because of Pull's working-class background.

Pull, not surprisingly, refused to let the report be published without his name on it. So the Worthing Archaeological Society simply refused to publish it, and they would not allow Pull to lecture on his work. What made Pull even more furious was that sometime later the society went ahead and published a full report on the excavations carried out by Pull and his friend C. E. Sainsbury, but under Curwen's name. Pull promptly resigned from the society. And he was not alone, for another who resigned with him in protest was the curator of archaeology at the Brighton Museum, Herbert Toms. He shared Pull's fascination for the folklore attached to fossil sea urchins in Sussex (see chapter 8).

The same year that Curwen was toiling away at Whitehawk, Pull, not daunted by the treatment he had received from his "fellow" archaeologists, had begun a new series of excavations near Blackpatch, at Church Hill in Sussex. It was here that he found even more fossil urchins that had first been collected in late Neolithic times. In association with a Beaker cremation in one of the mine shafts, Pull discovered three fossil urchins. He was convinced that these finds, and the ones he had found earlier at Blackpatch, indicated that these early people had consciously collected the urchins and buried them with their dead. He argued strongly for other archaeologists who excavated such burials "to note most carefully whether or not 'Shepherd's Crowns' [as the fossil was commonly called; see chapter 8] be present, either as single examples or in numbers; and if possible, to place the matter upon record." Pull was sure that these "fossil Echini . . . had some definite religious or magical significance."[31]

While Pull is only now finally receiving due recognition for his work, his death in 1960 brought him to national attention, though not for his archaeological endeavors. His brief notoriety arose from his untimely death at the hands of a bank robber. It was a sad end to a man who had been described as "one of the most gifted of amateur archaeologists."[32] His killer was fated to be one of the last people in England to be hanged.

Despite the animosity that existed between Curwen and Pull, what they both achieved was to highlight the fact that people in southern England during the Neolithic period were still, in some respects, hunter-gatherers—not of food, maybe, but of fossils. So, did these Neolithic fossil collectors, who wandered around the South Downs, as I did thousands of years later, simply share my fascination with these strange fossil urchins? Did the collecting instinct that runs so deeply through my veins form as much a part of the human psyche five thousand years ago as it does today? Or did my distant ancestors attach a much deeper significance to fossil sea urchins than I did, so important that they were sometimes buried with them—objects to accompany them on their last great journey, into the afterlife? Although it could always be argued that it was merely a coincidence that the urchins ended up in the graves, if these were just one-off chance associations of people and fossils, then this story would end abruptly here and now. But these few instances of people collecting fossil urchins are just the beginning. The story still has far to travel, through vast tracts of time and across the hazy distances of many ancient landscapes.

The hunt is . . . a distorted residue of what can still be recognized as a ritual that once formed part of a most ancient religious act. It is not true that the huntsman kills for the prize. That has never been the case, not even in prehistoric times, when hunting was one of the few ways to obtain food. The hunt was always surrounded by religious tribal ritual. The good huntsman was always the leader of his tribe and also in some fashion a priest. Over the course of time, all that has naturally faded, but even in their faded form, the rituals are still with us.

SÁNDOR MÁRAI, *Embers* (1942)

5

Fossilized Memories

Around twelve thousand years ago, as the Earth began to warm up after tens of thousands of years of being incarcerated in ice, people in the Middle East began to forsake a life of hunting, gathering, and fishing for a settled farming existence. Animals such as goats and sheep were domesticated, and a range of cereal crops were planted. This major change in lifestyle spread slowly northwestward across Europe from the "Near East," first through the Balkans and the Danube valley, before reaching northern France and the Low Countries about 7,400 years ago. Whether the farmers themselves moved, or just their ideas, it is not known. More likely, however, is that increased population pressure drove the change from a vagrant to a settled existence. But it was not until about six thousand years ago that farming people began to arrive in southern England by boat, carrying with them the seeds of new plants that they would grow as crops, and animals new to this relatively recently formed island.

The transition from a hunter-gather lifestyle to a settled agrarian existence seems to have happened incredibly quickly, in perhaps little more than a couple of hundred years. Radiocarbon evidence from the remains of cereals and domestic structures suggests that in a remarkably short period of time—perhaps less than two hundred years, between 4000 BC and 3800 BC—the more sedentary Neolithic way of life had become established across much of Britain.[1] These first Neolithic settlers brought with them not only new types of food, but also new tools to till the land. They also brought with them new ideas—ideas not only about how to live, but ideas about death, and of life after death.

Once these new cultivators of the land had settled on the downlands of southern England they began to clear the vegetation and work some of the chalky soils. Snail shells found in the ditches at Whitehawk indicate that about five thousand years ago the region was much wetter than it is now. This means that springs would have been abundant, flowing out of the chalk. Much of the downland would have been covered by trees and shrubs, such as beech, hazel, birch, and yew. As the thin, black soils were worked year in and year out and exposed to the rain, hard flint nodules concentrated at the surface as the finer material washed away. The more land that was cleared, the greater the potential for finding flints. Here was a more bountiful, readily obtainable source of material for tools than people had ever had before. And with more flints there came more fossils to be collected.

While the settled ways of farming reduced the need for a lifestyle exclusively occupied with hunting and gathering, the basic instinct to hunt was still deeply ingrained in the Neolithic psyche. So, hunting for good flints to work into tools, and fossils to satisfy the collecting desire, went some way toward assuaging the hunting urge. To pre-Neolithic people, the hunt seems to have been more than just a necessary quest for food. It was almost as much a religious act as a gastronomic one. How, then, could these breeders of sheep and growers of grain shake off a basic urge to hunt and gather, an urge that humans have carried deep within them for hundreds of thousands of years, and which was enmeshed with a spiritual ritual almost as intense as the kill itself? Now the hunter-gatherers' instinct was transferred to hunting and gathering other objects, such as fossil urchins, that would feed the spirit and the soul, rather than feed the body—the spiritual hunt had taken on a new dimension, and a new quarry.

As more flints were being picked up, rolled in the hand, and assessed for their use as tools by those who worked the land, more fossils would have been found. We know this in part because more fossil urchins (and other fossils) have turned up in archaeological sites of the Neolithic period than in older deposits. What would these Neolithic peoples have made of fossils? Like today, people thousands of years ago would probably have been attracted by the elegant, symmetrical shapes of fossil brachiopods and bivalves, and the mesmerizing spiral of ammonites and gastropods.

But more than just assuaging their hunger for the hunt, there is evidence to suggest that a range of fossils were considered to be objects of special spiritual significance. In a Neolithic passage tomb at Ballycarty in County Kerry in southwest Ireland, a large quantity of fossils was found that had been deliberately placed in the tomb.[2] These fossils included both straight and coiled nautiloid cephalopods, coiled goniatitic cephalopods, three different kinds of brachiopods, crinoid stems, bryozoans, and coiled gastropods. All collected from nearby Carboniferous limestones, these fossils had been placed along with an assortment of other grave goods, including bones of various animals, water-worn stones, a stone pendant, pottery, and pins. Each item, perhaps, was there to aid the departed soul in their life beyond life. And the recipients of these goods are represented now by just a few remnant burned bones.

In addition to collecting fossils like these, as well as fossil urchins in areas where they were plentiful, Neolithic people also seem to have been attracted to the spiraling fossil ammonites that are common in many parts of southern England. Outside a Neolithic long barrow at Stoney Littleton in Avon, one of the stone uprights that flank the entrance to the chambered gallery of this communal tomb has a mold of the dinner plate–sized ammonite *Arietites* cf. *bucklandi* impressed on its surface like a frozen, coiled snake.[3] Usually unnoticed but probably just as significant is the large counterpart slab on the other side of the entrance. This is covered by the fossil bivalve *Gryphaea*, long known locally as "devil's toenails." When I visited the barrow in the autumn of 2003 I found that it was not only our five-thousand-year-old ancestors who had placed fossils in the barrow. On its northeast corner is a large tabulate rock. The southeast corner had a counterpart. But on the one in the northeast corner, somebody had recently placed two fossil trigoniid bivalves—and beneath them a bunch of carnations. Old habits, it would seem, die hard.

Equally intriguing are the ammonites that have been found on Glastonbury Tor in Somerset. Excavations on the Tor by the archaeologist Philip Rahtz have revealed evidence for Neolithic and even earlier activity there in the form of axes, as well as flint and chert blades, material foreign to the region.[4] The Tor itself is terraced in the form of a series of complex spirals, and was first modified in Neolithic times. According to Rahtz, such stepped natural hills, and artificially constructed ones such as Silbury Hill that also has spiraled terraces, may be analogous to the ziggurats and pyramids found in Africa, the Americas, and Asia: structures that when ascended "bridged the space between earth and the sky, the sun and the 'heavenly' regions."[5] He suggests that the inspiration for the spiraled structure, which processions may have climbed, may have been the ammonites that are common in the Glastonbury area.[6] During excavations carried out atop the Tor in the 1960s, many hundreds of ammonites were found.[7] While some could have come from the Tor itself, the large number indicates that many were brought there by human agency. Rahtz concludes that "perhaps the spiral form of the ammonites had a ritual significance on the Tor related to the terraces; and perhaps not too distant from the whole 'meaning' of Neolithic art."[8]

Unlike the spiraling symmetry of the ammonite, it is likely that the smooth, elegant, round shape of the fossil sea urchin, so pleasant to cup in the palm of the hand, was part of the attraction the Neolithic people had to it. But so, too, was the five-rayed star that was so elegantly etched on its surface. It's hard to imagine what these people would have made of these objects that turn up in archaeological sites more often than other fossils. Would they have thought that their ancestors had created them in the distant past? Or did they invoke more ethereal, spiritual beings as the perpetrators?

Along with the recently developed techniques of farming that spawned a new generation of fossil collectors, these settled communities developed different ways of burying their dead. They constructed homes for them: passage tombs, like that at Ballycarty, where three circles of stones formed a series of burial chambers, to megalithic monuments such as barrows. Many still make an impact on the landscape of northwestern Europe thousands of years after their construction. The care taken over erecting these edifices for the dead is strongly suggestive of the development of complex spiritual beliefs associated with the transition from life to death.

Writing on the development of Neolithic cultures in Europe, Alasdair Whittle has argued that the dead were made to feel part of the living, probably as a way of denying the finality of death:

> The ancestors were invited to occupy stone houses, dark, quiet and difficult of access, and cajoled to remain with the hospitality of gifts of food and stone. In their honour, large upright megaliths were set up, trees of stone that resisted time and the seasons. In later generations, with the spirits firmly rooted and now taken for granted, other forms of stone shrine were built, holding the idea of ancestors but allowing easier access for the comings and goings of people. Some chambers in the stone houses of the dead

were roofed with pieces of earlier standing stones, fragments of memory from a timeless past. Human ancestors were laid in the shrines of the spirits. Now there were appeals to dry bones, and fingerings of empty skulls and motionless limbs. Human descent was traced through union with the spirits, and the shrines fostered regeneration, celebrated harmony in the universe, and expiated the guilt of beginning to domesticate the natural world.[9]

Along with being the builders of funeral mounds, these people were the creators of great stone structures like Stonehenge and of wooden temples like Woodhenge—and worshippers of unknown deities. But how can we be sure that even the placement of fossil urchins in the humble graves at Whitehawk carried some sort of spiritual significance? To discover how important these fossils must have been to Neolithic people we need to look elsewhere for similar practices—back to the place where these farming people had come from. Back across the English Channel to mainland Europe.

During the first two years of the Second World War numerous munitions factories were under construction in Germany. One was located at the foot of a mountain called the Euzenberg, near Duderstadt in Niedersachsen. During the course of this work traces of a dense Neolithic settlement were revealed.[10] For well back into the nineteenth century much of the land on the slopes of the Euzenberg had been owned by a single family, the Bertrams. This land they put over mainly to growing potatoes. Dig the soil, plant the seed potato. Dig the soil again a few months later and remove the potatoes. It's hard to estimate how many potatoes the Bertrams must have dug up, year after year, decade after decade. Tens of thousands? Hundreds of thousands? Maybe even millions? But every so often the soft, red, sandy soils yielded something quite unusual: strange potatoes as hard as stone, and marked with a star—fossil urchins preserved in flint. Although it wouldn't be very strange to find such fossils on the chalk downlands of southern England, it certainly was very odd to find them here, where the soil is underlain by rocks much older than the flint-bearing chalk. Fossil sea urchins just do not occur in these 220-million-year-old Triassic sandstones.

The unexpected presence of these fossils in an area where people lived in Neolithic times points to the fossils having been brought into the area many thousands of years earlier. Interestingly, all the urchins found here were the same type as the Whitehawk urchins—the helmet-shaped *Echinocorys*. The people who must have collected these fossils belonged to some of the earliest farming cultures on the North European Plain. With a way of life known as the Linear pottery culture, they had very rapidly colonized this part of Europe between 7,400 and 7,000 years ago.[11] Fossil sea urchins, such as those found at Euzenberg, that were not associated with a burial, but which were found as objects that are foreign to the region, and which can only have been brought in by humans for some unknown reason, provide another insight into the collecting mentality of Neolithic people. These fossils had been "hunted" in distant parts. They had been "gathered," collected, cherished, passed on, perhaps, from generation to generation, and taken far from where they first emerged from the ground.

The flint toolkits used by these people of the Linear pottery culture were based on the controlled production of blades, made from raw material that was often imported from far away. Trading in this valuable commodity must have been an important part of the society's interaction with other communities. In addition to bartering for the raw material to make their tools, they were also able to barter for other objects derived from the same source, including flint sea urchins. Unlike the flint used for blades, the Euzenberg fossil urchins seem to have served no apparent practical purpose. Their significance may have been more spiritual in nature. One possibility is that they formed part of a ritual associated with planting the seeds that would, it was hoped, turn into crops, ensuring the survival of the community. Maybe it was thought that by casting them into the ground, they would help ensure renewal of the crops. Whatever the reason, they must have been of sufficiently high significance to warrant acquiring and transporting over many miles.

Arguments supporting the view that fossil sea urchins had some sort of ritual significance in Neolithic times are supported by a number of other discoveries. These take different forms. Fossil sea urchins can be recognized as being in an archaeological context by their association with archaeological objects, such as artifacts. Or they may be placed in a particular context, such as a burial chamber or a grave. Like the graves in southern England, fossil urchins have been found in Neolithic graves in Denmark.[12]

The third form of discovery, as with the Paleolithic examples, is that many of these urchins have been found to have had been modified by humans to create functional items, such as an axe, hammerstone, or scraper. This modification may be of the fossil itself, or of a flint within which the fossil is set. For instance, in a Neolithic deposit at la Motte St. Jean, Saône-et-Loire in France, although no skeletal remains were uncovered, a polished axe was found along with three fossil sea urchins, all of a species called *Pegiocidaris coronata*.[13] Little doubt, then, that the urchins were first collected thousands of years ago. This type of urchin differs from forms such as *Micraster* and *Echinocorys* in that it is a "regular" urchin, one of Robert Hooke's "Button-stones," and circular in outline. The feature that makes one of the la Motte St. Jean urchins distinctive, and confirms that such urchins were collected for a specific purpose, is that it has been artificially perforated. The hole runs through the fossil urchin from what would have been its anus in life through to its mouth. Someone had also attempted to drill holes in the other two urchins, but had been unsuccessful.

Drilling holes in fossil urchins has a long history, particularly in France. Of the nearly two hundred examples of such fossils found in that country, about three-quarters come from the region of Saintonge in Charente-Maritime.[14] Although most of the perforated urchins from Saintonge probably date from Neolithic times, the practice seems to have continued for a few thousand years into the Iron Age. One reason why so many of these perforated regular urchins have turned up in archaeological sites in this region is because this type of urchin is commonly found in rocks in the district. However, even though they are also found very often in rocks in other parts of France, they are rarely found perforated in these other areas. This phenomenon points to a particular local cultural habit in the Saintonge district for perforating urchins. For some reason people were attracted to these objects, turning them into something they could use. Indeed, the habit appears to have been so strongly engrained in this society that it even extended to making artificial fossil urchins out of clay when there weren't enough of the real thing to go around. Similarly, in the Neolithic settlement of Oberjohnsdorf in the Reichenbach district in Germany, a clay spindle whorl was found that had been clearly fashioned after a fossil sea urchin.[15] As I discuss in chapter 13, this points to these fossils having as much a spiritual significance as a practical one.

The association of axes and sea urchins seen at la Motte St. Jean is, rather surprisingly, a recurring theme for thousands of years, from Neolithic through to Roman times in northern Europe. One such example comes from late first- to third-century remains of Gallo-Roman temples and wells at the Forêt de Rouvray near Rouen in France, where a cache of twenty small Neolithic axes was discovered, along with twenty-two fossil sea urchins.[16] These fossils were of the irregular urchins *Micraster*, *Echinocorys*, and another conical form called *Galerites*.

The reverence with which fossil urchins must have been held is shown in another way—by their presence in megalithic monuments such as barrows, those huge, artificially created mounds that often (though not always) contain the remains of the dead. Although the simple graves in which the Whitehawk skeletons were found do not seem to have been created with any great reverence, someone at least took the time to place fossil sea urchins with the bodies, expressing, perhaps, a modicum of concern for the spirits of the departed souls. Elsewhere greater care and respect were observed.

Barrows, though, were not always just mounds built over dead bodies. They also served a very important ritualistic role as receptacles of sacred offerings. Barrows ranging in age from the Neolithic to the Iron Age in France have yielded fossil urchins. At least two barrows have also been found to contain bronze bracelets, suggesting that the urchins may have served as some sort of amulet. Most surprising, and perhaps most significant, in at least three different sites in western France, lone urchins have been found in this type of barrow. Nothing else was inside, implying that these huge mounds, often tens of yards long, had, quite incredibly, been constructed merely to house the fossil sea urchins.

One such site is the Tumulus de la Fourcherie at Juick in Charente Maritime in western France, which is probably Neolithic in age. Within the barrow a single fossil urchin was found,[17] and nothing else—no bones, no ashes, no other object. Barrows that lack human burials have been called remembrance barrows. One of the most impressive is a small barrow about eleven yards (12 m) in diameter in Coatmocum-en-Brennelis, Finistère, in the westernmost part of Brittany. In this early Bronze Age barrow, constructed about 2000 BC, all that was found when it was excavated was a single fossil sea urchin protected by three flat stones.[18] Indication that the fossil was something very special is also shown by the fact that it was carried to its final resting place in the barrow from more than 125 miles (200 km) away from where it would have occurred naturally.

The Tumulus de Poiron, at Saint-Amand-sur-Sevres, Deux-Sevres in western France, is much the same. Even larger, being some twenty-two yards (20 m) in diameter and almost four and a half yards (4 m) high, this barrow was composed of layers of local schistose rocks surrounded by a circular earth mound. In the center a single fossil sea urchin had been placed in a box made of six shale tablets, about eight by three inches (20 by 8 cm).[19] Again, nothing else was contained within the barrow. These three barrows show very clearly that four thousand to five thousand years ago a deep symbolic meaning must have been attached to fossil sea urchins in northwest Europe.

Until relatively recently, there was a distinct difference between fossil urchins found in deposits associated with Paleolithic hunter-gathers and those found in the younger Neolithic ones. Hardly any fossil urchins collected in the more recent Neolithic times seem to have been adapted in some way as artifacts, like the Swanscombe hand axe. This suggests that maybe there had been a cultural shift in how people regarded these fossils as their lifestyle shifted from hunting and gathering to predominantly farming. More than five thousand years ago, as people in northern Europe settled into this sedentary farming existence, they began to realize that there was yet another source of flints. Other than just picking them up from their recently plowed fields, or from river gravels, as their Paleolithic forebears had done, they initiated an industry that involved active mining. By digging deep, vertical pits into the chalk they were able to extract high-quality flints. These Neolithic flint mines are known from many places in northern Europe, such as Spiennes in Belgium, Jablinnes and Serbonnes in France, Grimes Graves and Cissbury in England, and Kvarnby in southern Sweden. Studies of artifacts produced from the mines at Grimes Graves have suggested that the flints were not always being collected to produce tools for everyday use.[20] They were probably also being mined to obtain special ones that could be used solely for ceremonial purposes.

During their concentrated excavations underground for flints, it was almost inevitable that every so often Neolithic people would have come across fossil sea urchins, such as a long, thin flint nodule which has a domed *Echinocorys scutata* perched at one end.[21] This specimen has been recovered from the flint mines at Cissbury in Sussex. Radiocarbon dating of deer antlers used as picks to dig the shafts indicates that these diggings were made about 4,700 years ago. Interestingly, the paleoanthropologist Kenneth Oakley, who first described this object, thought that rather than having been derived from the mine itself, the flint capped by a fossil urchin had actually been brought into the mine. It does not appear to have been altered in any way. Noting its overtly phallic appearance, Oakley suggested that it may have been used as a flint fertility symbol. Perhaps it was brought into the mine to ensure an abundant supply of flint.[22]

For many years Oakley studied the archaeological occurrence of fossils, as well as the folklore attached to them. He firmly believed that people's fascination for fossils passed through various stages of cultural evolution, and as a result there was a link between the folklore attached to fossil urchins in recent times and the ancient spiritual or symbolic connotations. Oakley (1911–1981) was a paleoanthropologist at the British Museum (Natural History), and has many claims to fame. He is most renowned for his role in helping to unravel the Piltdown Hoax, one of the most famous frauds in the history of science. The Piltdown Man was the name given to some fragments of human skull and jaw, the first of which was found in Piltdown quarry in Sussex in 1912. Given the respectable scientific name *Eoanthropus dawsoni*, the species name being in honor of his finder, Charles Dawson, Piltdown Man was long touted as the "missing link" between apes and humans. In him the noble skull features of *Homo sapiens* were combined with the savage jaw of an ape. However, in 1953 Oakley and his colleagues published an article exposing the bones as fraudulent.[23] Less well known is Oakley's fascination with the folklore and symbolism of fossils, on which he wrote a number of articles. In these he not only documented many examples of fossils in folklore, but also recorded their occurrence in numerous archaeological sites.

Oakley's research led him to the conclusion that human fascination with fossils falls into three distinct phases. "Although," he wrote at the beginning of his article on the folklore of fossils published in the journal *Antiquity* in 1965,

> in the earliest phases of culture certain fossils were perhaps just regarded as "lucky," in more advanced phases they would be thought to contain magical power, and then, as animism gives place to belief in gods and ghosts, the fossil became a fetish, or habitat of a god; finally, when the religion decayed or was replaced by another the fetish was no longer an object of specific belief, but degenerated in folk memory to become regarded once again merely as an object conferring "good luck."[24]

As we shall see, the fossils may then pass into a fourth phase, where they become objects of mere curiosity that are collected for collecting's sake.

The Acheulian hand axe, and other Paleolithic examples of tools either made from fossil urchins or incorporating them, may all be considered as falling into Oakley's category of "lucky" objects. But the careful placement of fossil urchins in the Neolithic Whitehawk graves and in other barrows speaks of much greater significance. Were the fossils objects of great "magical power," or were they a "fetish, or habitat of a god"? Possibly some elements of the folklore that has persisted with these fossils into more recent times can tell us why these fossils, more than any other, appear to have been so important to many peoples, throughout Europe and beyond, for thousands of years (see chapters 8 and 9). Although folklore may be little more than a dim echo of much more significant attributes ascribed to these fossils, it may still help in the reconstruction of myths that they engendered.

In addition to the possible symbolic role of fossil sea urchins during Neolithic times, there is evidence from other sites that some were used in a more utilitarian manner. Even here, though, a secondary symbolism cannot be ruled out. At a Neolithic flint-knapping site at Sapinières in the Fôret de Montgeron in northern France, flint molds of *Echinocorys* and *Micraster* have been found that were consistently used as hammerstones.[25] They would have been relatively easy to hold in the hand, despite their relatively small size. But those who used them may also have been drawn to the distinctive five-rayed star on their surface. Users may also have been impressed by the starlike sparks that were spat into the air every time the fossils were struck on other flints.

Similarly, fossil urchins were collected during Neolithic times in the Umbria region of Italy. Of nineteen fossil urchins found by Giuseppe Bellucci in the last years of the nineteenth century, all, bar two, had been intentionally chipped.[26] Bellucci believed that this had been done to reveal whether or not the urchins had a silicified core. On the surface the fossils appeared to be made of much softer calcite, because they retained their calcite test (hard outer covering), but in many specimens this hid a silicified central core. Consequently, small flakes had been deliberately removed from either end of the elongate urchin by the Neolithic collectors so they could check the nature of the inside of the fossil. One of the other two urchins consisted or nothing more than the central core, which had been extensively chipped. So here the urchins were not being retained whole or turned into tools, but were being used in a more utilitarian manner to provide small flakes. But why choose the fossils? Could these Neolithic fossil collectors have believed that stones carrying a five-pointed star yielded flakes that carried with them more power than those knapped from more mundane stones?

The most spectacular demonstration of the continued attraction during Neolithic times of fossil urchins for those who worked flints and used flint tools has come to light in recent years, thanks to the energy of one man, the Belgian flint and fossil collector Roland Meuris. For a quarter of a century, Meuris has satisfied his collecting urge by scouring a small hill at Spiennes in Belgium called Camp-à-Cayaux for fossils and flint artifacts. Much of

Neolithic flint stone tool with fossil sea urchin *Micraster*, from Camp à Cayeux, Spiennes, Belgium; collected by Roland Meuris. Scraper length 3.9 inches (10 cm). Photo by D. Ceulemans, courtesy of Roland Meuris.

this area is heavily overgrown with bushes and small trees, or is pasture. But about sixty-two acres (25 ha) is a cornfield, and it is here, after each harvest, that Meuris searches for flint artifacts. Over the years he has built up a reasonable sized collection of flint tools as well as a large collection of fossils, especially sea urchins. But in 1984 his focus of collecting abruptly changed. That year he bought a copy of the Dutch edition of Bev Halstead's popular book on fossils called *In Search of the Past*. Leafing through it, he was suddenly transfixed by a photograph—it was the Swanscombe hand axe. Intrigued by the tool's combination of his two collecting passions—fossils and flint artifacts—and by the fact that nothing similar had ever been found at Spiennes, Meuris reasoned that Neolithic people might also have been inclined to produce tools incorporating fossil urchins. He set out to find one.

It took four years of hard searching before he finally did—a flint carefully knapped around a fossil urchin. Old habits, even of our Paleolithic ancestors, certainly die hard. After more than four hundred thousand years, their Neolithic descendants were still not averse to carefully manufacturing the odd flint implements around a fossil sea urchin. Since his first discovery, Meuris has found nearly one hundred fossil-urchin-bearing artifacts.[27] A far smaller number of artifacts (twenty-four in total) contain other fossils, specifically bivalves, brachiopods, and belemnites.

Some of the fossil-bearing artifacts are axes; others are scrapers or picks. In some the flint had been broken to reveal a cross section of an urchin. But in many the fossils are positioned away from the cutting or scraping edge and lie on the part of the tool that fits into the hand. In some instances it may be merely fortuitous that a fossil was on the piece of flint that was chosen to be worked. But in flints with a prominent urchin sticking out of the rock, it is inconceivable that the toolmakers would have been unaware of it. Everything points to them having worked the flint in such as way as to preserve the fossil. If you spotted either a flint with a fossil on it or an isolated fossil urchin, wouldn't you prefer to pick it up, rather than a humdrum piece of featureless flint? Some of Meuris's finds consisted of individual urchins that had been knapped on the edge and turned into scrapers, in much the same fashion as the urchin from Saint-Just-des-Marais that may be hundreds of thousands of years old. While other modern collectors might just discard chipped, damaged fossil urchins, Meuris has realized that other collectors, roaming this area five thousand years ago, may have had every good reason to intentionally break the fossils.

Any doubt that Meuris might have had that his artifacts were capable of being efficient cutting tools, he has most effectively dispelled. Occasionally he gives talks about his finds. A butcher by trade, he is often asked how our ancestors used the tools, and how good the flints were at cutting. He finally decided at one talk that there was only one way to convince his audience just how good the flints were. With all the panache of a magician he suddenly, much to the astonishment of his audience, pulled a recently deceased rabbit out of his bag. With a dazzling flourish of one of his Neolithic flint tools, the carcass was skinned, gutted, then cut it into pieces ready for the pot, all in the space of five minutes.

A hundred years earlier in Dunstable in southern England, another amateur archaeologist had been holding his audience similarly spellbound. But rather than creating the prelude to brandied rabbit in mustard sauce, he was using archaeological artifacts to prepare a loaf of bread. Worthington George Smith's party piece was to haul by horse and cart a number of stone querns, or hand mills, to the hall where he was to lecture. During the course of his lecture on how ancient humans made their bread, he would invite two members of the audience to step up onto the platform and spend much of his lecture hand-grinding wheat. The audience would then be served slices from a prebaked loaf made in a similar manner.[28] Apart from similarly enticing their audiences with gastronomic delights prepared in the ancient ways, Meuris and Smith had something else in common: they both had the knack of making spectacular finds of fossil sea urchins that had first been collected thousands of years before.

Just a year before serving up his gritty bread, Smith had made a most remarkable discovery in a field freshly sown with oats and turnips. Out of the chalky soil of a long-since-flattened Bronze Age burial mound on Dunstable Downs, to the north of London, Smith had excavated, piece by shattered piece, the skeleton of a young woman and a small child. Around the time of the discovery, in March 1887, the exhumation of prehistoric skeletons was not overly unusual. However, what made this particular find so different was that along with hundreds of fragments of fragile bones, Smith also unearthed, to his great surprise, a quite extraordinary number of fossil sea urchins.

Dead, long dead,
Long dead!
And my heart is a handful of dust,
And the wheels go over my head,
And my bones are shaken with pain,
For into a shallow grave they are thrust,
Only a yard beneath the street,
And the hoofs of the horses beat, beat,
The hoofs of the horses beat,
Beat into my scalp and my brain.

ALFRED, LORD TENNYSON, from *Maud*, part 2 (1855)

 # 6

Maud

It was a clear March morning in 1887 when Worthington George Smith set off from his home in High Street, Dunstable. Had you the chance to peek through the front door of Smith's house before it closed behind him, the eclectic nature of his interests would quickly have become apparent. The first room to the right off the hallway was his study. Like virtually every other room in the house, it was packed with a wild assortment of objects—from the antiquarian (including skeletons and cremation burials), to the literary (books and periodicals), to the artistic (finished and half-finished drawings and artist's paraphernalia).[1] These were all stuffed in boxes or in cabinets or piled on tables, chairs, or the floor. The study was dominated by a huge oak desk. Littered with half-finished drawings, piles of papers, and an assortment of pots, it so overshadowed everything else that it almost *was* the room. Peering down imperiously from the marble mantelpiece was the bust of a Roman, trying to make some small impact in this sea of oak, despite the fact that he was partly obscured by a large portable drawing stand. Stacked folios of sketches of fungi and tantalizing fragments of cathedrals, along with handwritten notes, covered what was left of the available floor space. There was precious little room for Smith to fit in; it was just as well that he was a slight figure. Regarded by his peers as a man of great character, with a fine brain and a good sense of humor, he stood out, even from a relatively early age, with his shock of white hair and well-developed beard. Inevitably dressed with dignified conventionality, Smith was a highly respected member of the community of the little Bedfordshire town of Dunstable.

Turning out of High Street and then left into Whipsnade Road, Smith began the long haul in his horse and trap early that morning on to Dunstable Downs. Perched high on the western edge of the Chiltern Downs, this area was well known for the many archaeological sites that scarred this chalk downland. As the horse slowly pulled his human load up the steep hill, the low morning Sun periodically flickered through the hawthorns and sloe trees bordering the side of the sunken road. The first primroses were beginning to flower in the banks on either side. The promise of spring was in the air.

Smith, now aged fifty-two, knew these hills well. Despite having settled in Dunstable only two years earlier, he had spent many long days roaming these rolling green downs. And it hadn't taken him long to come across evidence of the early inhabitants of the Dunstable region on these long rambles. Unlike today the grass was much shorter then, courtesy of hordes of rabbits, and aided and abetted by countless flocks of sheep that munched their way across the verdant swards. As Smith observed, "In the wettest weather wanderers might traverse miles of the breezy downs without getting wet-footed."[2] Short turf clearly showed up the slight hollows and ridges left by ancient settlements, whereas today they would be hidden by much longer grass. These irregularities were the remains of ancient hut circles. Smith's most impressive find had been a group of twenty-four circles, on the very scarp that he was now slowly climbing. But he had found little in the way of artifacts around them. And he had yet to find any skeletons. This made his excitement at the prospect of what lay at the top of the hill very understandable.

To his right as he broached the hill, Smith passed the cluster of low barrows called the Five Knolls Tumuli. But it was not to these ancient burial mounds that he was headed. His "quarry" lay on the other side of the road, where a seemingly much less dramatic ancient burial had just come to light. Smith's trap swung left onto a deeply rutted farm track, which soon meandered back to the north, almost in the direction from which he had come, trundling across bare, plowed fields that seemed to hold little promise of a rich harvest later in the year. The soil was little more than a mixture of chalk and flint. Smith never failed to be amazed by the magic that

Worthington G. Smith in his study. Photo by T. G. Hobbs, courtesy of Museums Luton/The Luton News.

farmers like his friend Frederick Fossey conjured up every year, in clothing the fields with such bountiful crops when the days were longer and warmer. And it was Fossey that Smith had to thank for getting him up here early that cold spring morning.

So what was it that made Smith, who had spent the better part of the last thirty years designing ecclesiastical drains and drawing fungi, want to head off to a cold, windswept, muddy downland field? For many years all that he had searched for when he wandered through woods or across open fields were fungi. It was something he did not just out of interest, but also to help pay the bills. As an accomplished artist his drawings of fungi helped just a little to feed his growing family. The first time that he had been paid for his illustrations of fungi, in 1867, he had received the princely sum of ten pounds for his first book. Entitled *Mushrooms and Toadstools: How to Distinguish the Difference between Edible and Poisonous Fungi*, Smith had been promised royalties from the sale of his book by the publisher, one Mr. Hardwicke. But Hardwicke had died, and the rights were transferred to another publisher and no more money ensued.[3] More galling was that a pirated version was published without Smith's knowledge or consent. Yet when he asked the publisher if he could revise his own book, he was told that it wasn't necessary. Even so, his little book was something that Smith was very proud of, despite the lack of pecuniary return over the last twenty years.

Arguably Smith (1835–1917) was one of the foremost botanical illustrators of the day, despite his lack of any formal training in drawing of any sort. And when it came to fungi nobody could touch him. He had been told this enough times by his fellow members of the Woolhope Naturalists' Field Club, founded in 1851 "for the practical study in all its branches of the natural history of Herefordshire." Smith joined in 1869. Despite the long trek from Islington in London, where he lived at that time, and then the equally long journey from Dunstable, he became a leading member. The club met at The Green Dragon in Hereford in the autumn, and one of the highlights was the Fungus Forays. On these expeditions, which often went on for a week, members scoured the woods for fungi. These were then collected and classified before being eaten each night at some comfortable hostelry. A chef was hired especially for the occasion. A contemporary report in the *Daily News* states:

> The members freely partook of several varieties [of fungi], not merely snatching a fearful joy from the repast, but deliberately criticising its characteristic features. Almost all Agarics [a particular form of fungus], it must be remembered, are rather rich, and should therefore be eaten in moderation. To eat them as accompaniments to meat is an obvious error; they should form a separate course and Burgundy will be found the most suitable wine to drink with them. With regard to their cooking it may be laid down as a maxim that they should never be boiled, but either stewed gently and for a long time, or broiled with salt and pepper before a quick fire.[4]

It had been more than twenty years since Smith had drifted into this rather esoteric pursuit of drawing (and eating) a wide variety of fungi. In all probability this arose as an antidote to the day-to-day activities that had previously brought him money to pay the bills—the design of somewhat more prosaic objects, namely cathedral drains. Even though these soaring edifices of stone which resonate to the glorious sounds of the Masses of Palestrina may be the pinnacle of our architectural achievements over the centuries, cathedrals also needed drains. But to an ambitious young man who had dreamed of designing the most splendid cathedrals in the land, and who had only been able to obtain work for a firm that specialized in repairing dilapidated ecclesiastical buildings and sorting out their drains, it was all too much. Even his stint designing ecclesiastical furniture for the

Roman Catholic Church, such as screens, altar tabernacles, and crosiers for the likes of Cardinal Wiseman and Pope Pius IX, was a period of artistic frustration.

Smith had started off on a much more positive note. When just nineteen years old, he enthusiastically suggested to Sir George Gilbert Scott, Surveyor to the gently decaying Westminster Abbey, that it might be prudent if Scott consulted "some eminent chemist about means to prevent the continuing decay of the stone, and if commended, that the means be tried on the pinnacles and parapets."[5] We can probably thank Smith that the Abbey still stands today, because three months after his suggestion the Chapter of the Abbey recommended Smith's application for preserving the stone on a trial basis on one of the towers. From that time on, the stone of the Abbey was treated and preserved.

Probably, when Smith thought back to those frustrating times, he felt that he was getting nowhere despite being full of ideas. When just twenty-six years old he had finally revolted from the drudgery of drain design. Thus began the rather precarious existence of trying to support himself, his wife, Henrietta, and the three of his seven children who had survived the ravages of infant illnesses by becoming a freelance illustrator and engraver, of both architectural and botanical objects. Much of Smith's early work in this field still had an ecclesiastical flavor. Somehow he managed the switch from architectural draftsman to botanical illustrator with little change in style. E. C. Large, writing in 1940 in his book *The Advance of the Fungi*, noted that on the occasion of Smith's having to draw a tomato in a hurry to illustrate an advertisement in *The Gardeners' Chronicle*, the end result "looked exactly like an ornament off a wrought-iron gate, and so realistic you could almost lift it off the page with a pair of tongs."[6]

But to keep bread on the table he produced copious freelance drawings for *The Builder*. Each week he was sent sets of architects' plans, and each week he transformed them into imaginative perspective views of completed buildings. Among these were his drawings of Scott's restoration of Westminster Abbey, and the interior of Cologne and Worcester cathedrals. For the next twenty years this work provided Smith and his family with a steady (albeit frugal) income.

The love of his life at this time, though, was fungi. But sometimes his gastronomic passion for consuming the objects of his artistic endeavors had led to near-fatal consequences. Ever one for practical experimentation, Smith, by personal experience, had discovered the poisonous nature of *Agaricus fertilis*. After collecting some of the fungi he kept them under a propagating glass for a couple of days. He then cooked them and fed them not only to himself, but also to his wife and two-year-old daughter, Edith. His son Alban had died in 1858 without having seen his first birthday. His daughter Agnes had died at birth. Another son, Stephen, had died just two years earlier at the same age that Edith was now. It therefore seems to have been an act of culpable lunacy to feed this fungus to his surviving daughter. Fortunately, she and his wife, along with Smith, all survived the ordeal, but not until after all three had become very ill, suffering from vomiting and dizziness. Smith, who had eaten most of the fungi, took fully two weeks to recover. They were lucky, because *Entoloma lividum*, as the species is known today, has killed infants.

Smith's rather cavalier attitude toward the edibility of fungi is also shown by another near-fatal consumption of a fungus. Writing to a friend in 1871, he recounted how he

> had some specimens of the fungi sent me that poisoned the family at Falmouth. They were abnormally grown, strange scented specimens of *Maramius oreades* (the fairy ring champignon). I at once had six or seven cooked, and I ate them all. They were terribly hot in the throat afterwards and disordered my stomach a little. I cannot help thinking that this species is sometimes dangerous, for twice before it has upset me, and once as recorded in my little book it made me very ill, but I then thought some other species had got into the frying pan.[7]

And into the fire perhaps.

It is hard to decide whether Smith was just recklessly foolhardy or a dedicated gastronomic pioneer. Having decided to write a book on edible and poisonous fungi, he had been determined to find out for himself whether or not they could be eaten. But, being a man of many parts, he had other passions also. As it was to transpire, a knock on the door that had got Smith up early that March morning had nothing whatsoever to do with fungi. What had sent a clenched fist to rat-a-tat on his front door and shatter the early morning peace was a body—a body that, until that morning, had lain peacefully undisturbed for thousands of years.

The other great passion in Smith's life was investigating the remains of ancient Britons. Having developed the habit of scanning the ground for fungi when he was out walking, and being of a generally inquisitive nature and possessing the artist's eye for form, Smith had, some nine years earlier, found his first flint implement. He had been bitten by the collecting bug. At that time Smith was living in Shoreditch in north London, and terrace houses were spreading like a rather nasty rash across nearby Stoke Newington Common. To make the new roads along which the houses were strung, huge quantities of flint gravel had been brought in by rail and river from quarries near the River Lea at Hertford and dumped in great heaps on the Common. It was while browsing over some of these mounds of flint that Smith found his first Paleolithic flint implement.

What also inspired Smith had been the publication of one of the most influential books on archaeology to come out of the nineteenth century: Sir John Evans's *Ancient Stone Implements of Great Britain* (1872). Its description of two hand axes, one discovered in Hackney Downs and another in Highbury New Park, particularly excited his attention. Smith realized that there might be other hand axes closer to home. So for five years he, as he was to write later, diligently kept "a record of the exposed surfaces of every drain, house foundation pit etc." But rather than designing drains he was up to his neck in carefully hand-dug foundation trenches and drains being put in for new houses. What Smith found, exposed in the walls of these ditches, was a gravelly layer up to six inches (15 cm) thick in which he soon discovered many flint implements, countless flint flakes, antlers, and even two four-foot-long (1.2-m) birch stakes that had been sharpened at one end. As he was to record some years later in his own best-selling book, *Man the Primeval Savage*,

> I first noticed this thin stratum of flint, in some places full of Palaeolithic implements and flakes . . . in excavations on the south side of Stoke Newington Common; later in the spring I observed a similar stratum in the fields and market-gardens of the north side of the Common.[8]

Buoyed by his discovery of flint tools around London, Smith started looking for artifacts in some of the brick pits in the Dunstable area after he moved there. In his assessment of the significance of the Paleolithic sites in the Chilterns that Smith had collected from, Mark White notes how Smith's "tenacity and dedication were rewarded by the discovery of several outstanding earlier Palaeolithic sites . . . each yielding a significant, primary-context Acheulian assemblage."[9] The significance of his discoveries is such that as late as the 1960s, of the five known sites in Britain that had yielded Acheulian artifacts, four had been found by Smith. Strangely, there was little follow-up to his work for almost a century, the attitude long having been that little of value had been found by Victorian antiquarians such as Smith. But as White has pointed out, such diligent workers "deserve homage for the calibre and rigour of their work."[10]

Along with finding a range of Acheulian tools, Smith discovered what he believed was other evidence of the creative ability of the early Paleolithic peoples who lived in this area hundreds of thousands of years ago. He uncovered over two hundred spherical fossil sponges called *Porosphaera globularis* from Paleolithic deposits at Bedford. Looking disturbingly like flint chocolate Malteser candies, a distinctive feature of many of these sponges is the presence of a hole running through them. While this hole is generally a natural feature formed in all likelihood by the sponge growing around an elongate object, such as a stalk of seaweed, Smith thought that many appeared to have been artificially enlarged.[11] He also found one containing organic material that he believed represented fiber on which the sponges had been thread, leading him to suggest they had been used as necklaces. Recent reexamination of these and similar specimens from other archaeological sites supports Smith's view that these specimens were paleoethnological. Wear facets on some of the sponges are attributed to the erosion caused by adjoining sponges on a thread rubbing against each other.[12]

Whether Worthington Smith ever took his family on a trip to Brighton we do not know. But given the town's popularity in the nineteenth century, this wouldn't have been too surprising. One of the many differences between the late nineteenth century and today is that when "strolling" over the pebbly beach, the Victorian vacationer would have stood a good chance of being accosted by fisherwomen. One of their ways of encouraging benevolent tourists to part with their money was by selling them necklaces made from the selfsame fossil *Porosphaera* sponges that Smith had found in the Chiltern brick pits. These, though, had been picked up from among the flinty pebbles of the beach.[13]

And it was in 1887, around the time he started making these Paleolithic discoveries, that Smith became involved in the discovery of a much younger but nevertheless equally significant find. On that March morning, Smith was heading for the former site of a pair of round barrows, high on the downs overlooking the town of Dunstable. He "remembered them as a boy when they were high grass-grown knolls, standing about 10 feet high and 45 feet in diameter."[14] In a letter that he wrote to local naturalist Jannion Steele Elliot on July 22, 1913, Smith recounted how the graves were discovered:

> When the ground was brought under cultivation in the 1850s the knolls were almost levelled and an engine, known to farmers as a "steam cultivator" traversed the sites of the two knolls. At the present time their position

is only indicated by a very slight swell in the ground and the presence of a considerable amount of chalk rubble in their immediate neighbourhood.

A farmer friend of mine (F. T. Fossey), told two men to lower the ground. Directly the men began, they pecked into bones and took a few pieces out. The farmer stopped them and communicated with me. I then took over the job and had all the bones found. I was on the downs for two days, as the original size of the tumulus was large. I had my meals, beginning with breakfast, at the farmer's house. As there was another ruined tumulus on the same farmer's land, I was allowed to dig that over and a boy's skeleton came to hand, with old pottery etc.[15]

After excavating the skeletons for a couple of days, he felt he had become familiar enough with the first one to give it a name. During the excavation Smith realized that the skeleton was that of a female. So she was christened Maud. Why Maud? Perhaps Smith was familiar with Tennyson's long poem *Maud*: "Dead, long dead, / Long dead! / And my heart is a handful of dust . . ."

It didn't take him long to realize that the grave contained things other than bones—it also held a large quantity of fossil sea urchins. By the time all the bones had been removed and carefully packed in boxes, Smith and his helpers had found twelve fossil urchins. All had clearly lain close to Maud's body. Some were heart-shaped. Smith figured them later in his book, *Man, the Primeval Savage*, and called them *Micraster coranguinum*. The group also found some helmet-shaped urchins. These he called *Ananchytes ovatus*. Today this species is known as *Echinocorys scutata*.

But this was not the end of the surprises that Maud had in store for her exhumers. As Smith and his helpers extended the grave even further, more and more fossil urchins came to light. Barely a minute seemed to pass without a fossil being found. By the time they had stopped they had amassed, amazingly, almost another one hundred of the fossils. Yet this was still not the end. As Smith moved further away from the grave and raked over the chalky soil that had once covered the entire barrow, he entered what must have seemed a surreal world, where he was unexpectedly harvesting one of Mr. Fossey's better potato crops. But these potatoes were definitely not edible—they were made of cold, hard stone, and seared with a star. Dozens upon dozens of these fossil urchins emerged from the ground as he raked. The more he raked, the more he found: "On repeatedly shoveling and raking over the earth from the entire tumulus [which had contained a number of graves], 200 or more were found and most of these, undoubtedly, originally belonged to the girl's grave, as none were found in the other graves."[16] It is not clear whether the 200 included the original 112 or represented an additional 200, making a total in excess of 300. Either way it is an extremely large number of these fossils to be found associated with one burial.

In his excellent biography of Smith, James Dyer records fewer fossils as having been found. Following the discovery of the initial 12, he describes Smith as finding "another 91, making 103 in all," and that these came from "the earth that was thrown out of the entire barrow."[17] Interestingly Smith, a meticulous and careful artist, depicts 147 fossil urchins in his engraving of the skeleton. He was convinced that all the fossils had once lain in Maud's grave, for none of the other graves showed any signs of having contained them.

Later that year, on Thursday, December 8, when giving a lecture to the Congregational Mutual Improvement Society of Dunstable, Smith was to report that there had been more than a single grave under the barrow:

> The excavations showed that the first barrow originally covered seven or eight graves. Of these the chief was a large central grave about 12 feet long and 6 feet wide. This central grave had been dug through the surface soil (of about 1 foot) and then carried 2 feet 6 inches deeper into the solid chalk. The central grave was entirely cleared out by our workman, but nothing of moment was found, for it had been previously dug out, and a large modern iron nail was found at the bottom of the grave. Round this grave were six or seven other graves, each about 3 feet deep from the surface; all these with the exception of two had been previously disturbed, and the contents (whatever they may have been) thrown out and destroyed. One grave on the east side of the mound had not been previously opened, although the former diggers had been so close to it with their picks and shovels that they had seriously damaged the lower bones of the legs of the occupants. Several persons had probably been interred in the large and deep central grave, but the graves belonging to the circumference were all comparatively shallow and small; they were irregular in shape and measured roughly about 3 or 4 feet square. Each person must have been buried on his or her side with the knees drawn up under the chin.[18]

It was in one of these small outer graves that Maud's skeleton had been discovered. By the end of their diggings that day, Smith and his helpers had recovered 340 pieces of bone as well as the hundreds of fossils. However, as Smith was to discover later, Maud was not alone in her grave.

Smith's discovery was really quite astounding, although he probably

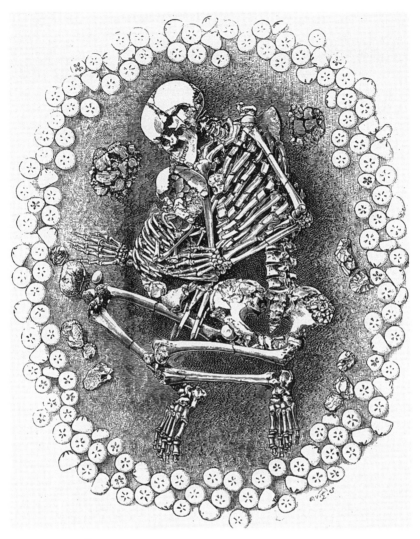

Skeletons of "Maud" and her child found in a Bronze Age grave on Dunstable Downs, buried with hundreds of fossil sea urchins. From frontispiece in Smith (1894).

didn't realize the full extent of its significance at the time. Never before (and, as it was to transpire, never since) had so many fossil urchins been discovered in a single grave. Along with the mass of fossils in Maud's grave, Smith and his helpers also recovered a few other items. As Smith catalogued them later in his book, they included "a broken and partly pulverised British pot, . . . Three globular hammerstones of flint, . . . two scrapers, two very rudely chipped celts, and a large number of flint flakes."[19] Very mundane fare, though, when compared with the paleontological riches that the grave had also contained. It was a fossil collection that even today any paleontologist would dearly love to possess.

Smith stayed on for another day, as Fossey knew of another grave in a nearby barrow that he thought Smith might like to excavate. Given the success of Maud's excavation, he clearly did. However, this second grave, probably much to their relief, was free of fossil urchins, but it did contain the skeleton of a boy aged (Smith thought) between about fourteen and sixteen years. These bones even more fragile than Maud's: "He was as flat as a pancake in the ground and in 445 pieces, every bone fragment being as soft as touchwood," Smith explained in his lecture later that year.[20] However, they were dutifully collected and carefully packed away, ready for transport back to Dunstable.

Some days later, after Smith and his skeletal friends were safely back home, he began the long task of carefully unwrapping all the bones from Maud's grave, prior to conserving and strengthening them. As he was to tell his spellbound audience later in the year, "Every piece was as soft as gingerbread. Each had to be washed and then thoroughly dried. They were then dipped in thin, very hot, almost boiling, gelatine and dried a second time. When the bones had become once more perfectly hard I sorted them and glued all the broken pieces, then fitted them together with thin glue."[21]

Smith's lectures were never dull. When he wasn't getting the audiences involved in helping him make bread or, as on one occasion, demonstrating ancient musical instruments, he was holding them spellbound with his vivid account of the discovery of Maud. It is not too difficult to imagine the scene. The only sound in the large hall is the constant hiss of the gas lamps, sputtering their sickly yellow light across the skeletons that he has spread out on the table in front of him. The audience is as deathly quiet as the bones. "It was just as I was first laying out Maud," Smith explains, "that I made a rather unexpected discovery. Almost from the first moment I saw her I was sure she was a young woman, mainly, I suppose, from the delicate features of her face." Smith stops, then slowly leans over the table and picks up a small, delicate, curved piece of bone.

He holds the bone up for his entranced audience to see. "Maud," he continues, "had not lain alone in her grave for thousands of years. Buried with her was a small child."

A soft murmur blows over the audience, like the susurration of wind across a wheat field. Those at the front crane forward to get a good look at this fragile remnant of what must have been a very young child. "I suspect that the child was no more than two years old," Smith adds.

"How either of them met their deaths is a mystery," he continues. "My feeling is that the child was buried alive with the woman, whom I can only presume was its mother."

Smith never elaborated on why he thought this was so. Whether he was thinking of a ritualistic human sacrifice, either of the child itself or of mother and child, we do not know. Yet there are some very interesting similarities between the circumstances of this burial and the one at Whitehawk. In both cases the skeleton was female. In both cases the women had died in their early twenties. In both cases a child was also present. In both cases fossil urchins were among the meager variety of grave goods.

It is unlikely, though, that the two women were living at the same time. The only estimation of when Maud lived is a suggestion of Smith's. He based his view on the very simple premise that her head was brachycephalic. What this means is that she had a relatively broad skull. In the late nineteenth century great emphasis was placed on whether skulls were either "brachycephalic," where the width of the braincase was greater than 80 percent of the height, or "dolichocephalic," where this ratio was less than 75 percent. At the time it was believed that Neolithic peoples in England had possessed the narrow, dolichocephalic skulls, but that the Celts who succeeded them and lived during the Bronze Age were brachycephalic. Because Maud fell into this second category, Smith concluded that she must have been of the Bronze Age type, Celtic stock.

It has since been realized that within any population of any racial group there is a large degree of variation, from people with narrow skulls through to broad-skull individuals, and with everything in between. Another indication of when Maud lived comes from the shape of the barrow in which she was buried and the method of her burial. Her barrow was circular. Such mounds characterize the early Bronze Age. Moreover, her remains were not cremated. Early Bronze Age bodies were almost invariably interred, whereas later Bronze Age bodies were usually cremated.[22] This indicates that Maud and her child were probably buried with their magnificent fossil collection about four thousand years ago.

While today we may see the five-pointed pattern on fossil urchins as symbolic of a star, would Maud have seen it the same way? The discovery in 1999 of a Bronze Age sky disc suggests that she probably didn't. Found at Mittelberg near Nebra in Germany, the bronze disc, the size of a large dinner plate, was crafted not long after Maud and her child had been laid to rest. What makes this disc so special is that it is thought to represent the first archaeological evidence for an undoubted representation of the cosmos.[23] On the disc are set what appear to be either a golden Sun or Moon, lying alongside a gold crescentic Moon. Two curved gold bands run around the rim (originally there were three). One of the gold bands and the missing counterpart on the edge of the disc are thought to represent the horizons. The ends of the gold bands subtend an angle of 82.5 degrees, which is the distance between the northernmost and southernmost points of sunset and sunrise in the area. The third curved gold band is set between the two horizon arcs. Tiny lines on the upper and lower edges may symbolize oars. If so, the band represents a ship coursing between the horizons across the nocturnal celestial ocean. Yet of all the gold symbols hammered onto the disc, perhaps the most telling are the thirty-two small circles that are scattered across the bronze cosmos. These almost certainly represent stars. Support for this interpretation comes from one particular cluster of seven circles that are thought to be a depiction of the constellation Pleiades.

If this disc truly reflects how Bronze Age people in northern Europe saw the heavenly bodies, then they did not see stars in the night sky as five-pointed or even six-, seven-, or eight-pointed images, but as simple circles of light. Whether or not the five-rayed pattern on the hundreds of fossil urchins that were buried with Maud four thousand years ago had a direct link to the stars, they may have played some part in rituals associated with her passage into the afterlife.

After Smith had shown Maud and her child to the current inhabitants of Dunstable, he had one more significant task to achieve—to draw the two skeletons, much as they would have looked in their grave. There can be little doubt that, despite being separated so far from them in years, Smith was very touched by his finding of the child interred in the grave with the woman. We know this not from his writings but from his drawings, particularly from the manner in which he drew what he had found. Smith was renowned for the very realistic style of his drawings. Indeed, Wilfred Blunt, writing in 1950, was critical of this very realism. "In the work of Worthington Smith," he wrote, "we reach the zenith of distasteful efficiency. His drawings are probably all that a botanist could desire, but to the artist they reveal little trace of sensitivity; the outlines are crude and monotonous, the shading mechanical and the general effect repellent."[24] Blunt clearly had never seen Smith's drawing of Maud and her child.

With this drawing Smith indulged in what was for him a great degree of artistic license. No straightforward, almost photographic representation of a pile of crumbling, jumbled bones here, peppered with hundreds of fossil urchins. On the contrary, what he produced is almost Madonna and Child–like in its composition. This gentle drawing shows the skeleton of the woman in the crouched position in which she was found. Nestling into her breast is the child, lovingly embraced by Maud, with her chin resting lightly on top of the child's head. Together they are wreathed in hundreds of fossil urchins, so densely arranged that they seem to be forming a barrier, protecting them from whatever they were to experience in their long journey into the afterlife.

When Smith's book, *Man, the Primeval Savage*, was published six years after he found their remains, this was the drawing that he chose to use as the frontispiece. The caption simply reads, "Skeleton of woman and child found in a round tumulus near Dunstable."

> An idea which has once been developed and found to serve its purpose is not lightly thrown away. It lives on persistently, though sometimes changed and obscured, and is spread far and wide. Entirely fresh and independent ideas are but seldom brought forward: one generation lives by the thoughts of those who have gone before. The idea itself and its outward expression pass on from people to people, often only with slight modifications suited to the special needs of the different nations.
>
> CHRISTIAN BLINKENBERG, *The Thunderweapon in Religion and Folklore* (1911)

7

Augustus Henry Lane Fox Pitt-Rivers

During the latter part of the nineteenth century the study of archaeology progressed in enormous leaps and bounds. An increasing number of more thoughtful amateur archaeological sleuths like Worthington Smith had become concerned with interpreting the significance of their finds in a reasoned and intellectual manner. They were seeking something more than just the simple desire of satisfying the collectors' urge to accumulate the buried treasures of antiquity in ever-increasing hoards. This is not to say that they weren't collectors. Certainly men like Smith were collectors. He, for one, undoubtedly had the collectors' passion. Indeed, he was a collector par excellence, having a house crammed from floor to ceiling with his archaeological treasures. But in addition to satisfying their innate acquisitional urges, this new breed of amateur archaeologists wanted to understand what these miraculous artifacts could tell them about life in distant times, and what they could learn about the people who had crafted them.

Dramatic developments had also been taking place in the study of geology during the nineteenth century. By the 1880s, when Smith was digging his way down through layers of gravel searching for flint tools, and into chalk to unearth Maud, her child, and their shroud of fossil urchins, geology had become a well-respected science. However, at the start of the nineteenth century it didn't even exist as a recognizable discipline. One man, though, changed all that—another W. Smith. This one, William Smith, created the first geological map of England and Wales, published in 1815. This Smith (unrelated, as far as we know, to Worthington Smith) was able to achieve this remarkable feat because he had reasoned that the relative ages of the strata that he saw in railway cuttings, in newly excavated canals, and in quarries that pock-marked the countryside could be ascertained based on the sequences of fossils that they contained. This was the key to unlocking the mysteries of the ages of the limestones, sandstones, and shales that molded the landscape. Once a scheme had been developed that enabled two outcrops of rocks many miles apart to be correlated, then a geological map could be constructed.

As a young boy living on a dairy farm in Oxfordshire, the objects that had most fascinated William Smith were circular stones, four to five inches (10–13 cm) across and with flattened bottoms and curved tops. According to his nephew, John Phillips, Smith knew these fossils as either "poundstones" or "quoitstones."[1] (Phillips also recorded that Smith's uncle "was very little pleased with his nephew's love of collecting.")[2] Commonly found in the district, they were called poundstones because sometimes they were used as weights on butter scales. What made them useful for this purpose was not only their amenable shape, but that they all weighed much the same. However, from a report written by Smith in 1804 concerning field observations he had made in 1790, he preferred to call them quoitstones rather than poundstones.[3]

These fossils were given quite a different name in the seventeenth century. Writing in his *Natural History of Oxfordshire* (1677), Robert Plot observed how the shape of these fossils

> gave occasion to a Learned Society of *Virtuosi* [the Royal Society], that during the time of the late *Usurpation* [the English Civil War] lived obscurely at Tangley, and had then time to think of so mean a subject, by consent to term it the *Polar-stone*, having ingeniously found out by claping two of them together, that they made up a *Globe*, with a *Meridianus* descending to the *Horizon*, and the *Pole* elevated, very nearly corresponding to the real *elevation* of the *Pole* of the place where the *stones* were found.[4]

But their other special feature was the five-rayed star etched onto their upper surface—for these stones were fossil sea urchins. These urchins, now called *Clypeus plotii*, were much larger than the urchins from the chalk that had been placed in Maud's grave. They were also far older. More than 100 million years older, in fact, because these large urchins had lived during the Jurassic period.

As Smith learned more and more about the other fossils that he found in the yellow Jurassic limestones and grey mudstones of the Cotswolds, including brachiopods, bivalves, and ammonites, he began to realize that strata at different levels contained different types of fossils. Strata close together would yield fossils that were most similar. Strata much higher, or much lower, on the other hand, would contain suites of fossils that were more different. Moreover, Smith reasoned that of two strata, one lying on top of the other, the lower will be older than the one on top. And so was born not only the science of geology, but also the realization that rocks were nature's clocks. The world, at the beginning of the nineteenth century, had entered the fourth dimension.

If Charles Darwin was the "Father of Modern Biology," then William Smith was the "Father of Geology." Smith had presented the world with the unassailable realization that the Earth must be far older than the six thousand years espoused by the church. The discovery that layers of rock could be read like a book (a book written over what then must have seemed to be an unimaginable period of time) undoubtedly paved the way for naturalists like Darwin and Alfred Wallace. If their ideas of evolution by natural selection had any basis whatsoever, they would need more than the biblical six thousand years to explain the myriad life-forms existing on Earth, and the countless numbers whose fossilized remains peppered the rocks. So it was that the publication in 1859 of Charles Darwin's ideas on the origin of species by natural selection,[5] some forty-four years after Smith's map was first published, led to a welcome burgeoning in attempts to unravel the mysteries of the past—to generate new ideas, bury some old ones, and explore and modify others.

A number of paleontologists in the late nineteenth century, like Arthur Rowe, who documented evolution in species of the heart urchin *Micraster* in the chalk downland, were attempting to apply Darwinian ideas to their field of study.[6] But so, too, were a few archaeologists. Foremost among this new breed of meticulous, questioning archaeologists who followed the principles of both Smith and Darwin was Augustus Henry Lane Fox Pitt-Rivers (1827–1900). Here was yet another pioneer in his discipline, for many people regard him as the father of modern scientific archaeology. He was renowned for carrying out his excavations in a methodical, disciplined, and detailed way, and unlike many of his contemporaries, he interpreted the significance of what he found.

Pitt-Rivers was born as Augustus Henry Lane Fox in Yorkshire in 1827. At the age of fourteen he entered the Royal Military Academy at Sandhurst and at eighteen was commissioned into the Grenadier Guards. He fought briefly, but with distinction, in the Crimean War. Then he was on the staff at the Battle of Alma and at the Siege of Sebastopol, where he was mentioned in dispatches. Pitt-Rivers also served in Malta, England, Canada, and Ireland.[7] Early on in his career he developed a professional interest in the development of firearms through the ages. This arose from his early membership in the committee on the adoption of the rifle into the British army, and subsequent position as first instructor of the Hythe School of Musketry.[8] He began to collect many varieties of both offensive and defensive weaponry and accessories, in the process accruing an appreciable collection of many different varieties of ethnographic objects. As Sir Edward Burnett Tylor, first Professor of Anthropology at Oxford University, described it,

> In order to follow . . . (the evolution of design in firearms) he collected series of weapons . . . the method of development series extending itself as appropriate generally to implements, appliances and products of human life, such as boats, looms, dress, musical instruments, magical and religious symbols, artistic decoration and writing, the collection reached the dimension of a museum.[9]

These objects Pitt-Rivers kept at home, where they lined the walls of his London house from cellar to attic.

Whether it was his military service that taught him to excavate methodically and developed his keen eye for detail, it was a nonmilitary influence that, in many ways, guided Pitt-Rivers's later work. He had been highly influenced by Darwin's ideas on how species of animals and plants evolved. To his mind, the natural thing to do was to try and explain material cultural development in similar terms: the transformation by cultural selection of the fitter, better adapted forms from the simpler, more primitive ancestral forms. He used firearms in his first attempts to fit Darwin's ideas to aspects of the cultural world. On the basis of these, and later other ethnographic material that he had accumulated, Pitt-Rivers introduced the term *typology* to explain his method of arranging artifacts in a chronological and developmental sequence.[10] His view was that, like organic evolution, material culture similarly evolves from generation to generation. As his interest in older artifacts took him deeper into the world of archaeology, Pitt-Rivers

Portrait of Augustus Henry Lane Fox Pitt-Rivers (PRM 1998.271.6).
Photo courtesy of the Pitt Rivers Museum, University of Oxford.

was driven by the desire to come up with a scheme that would show the historical evolution of cultures, based on evidence drawn from artifacts. His ideas were formulated by analogy to the philosopher Herbert Spencer's theory of social Darwinism, a concept that Darwin himself dismissed.[11] Spencer's ideas were based on the premise that evolution necessarily equates with progress.

Pitt-Rivers's posting to southern Ireland in the 1860s appears to have stimulated his interest in archaeology, and he spent time surveying some prehistoric sites.[12] In 1864 he was elected to the Society of Antiquaries, and his contacts in the archaeological world blossomed. This wide range of experts was not only able and willing to assist him with the acquisition of future artifacts, but they could advise him on objects already in his collection.

After spending his military years traveling and acquiring a sizable ethnographic and archaeological collection that numbered about fourteen thousand items, in 1880 Pitt-Rivers was placed in the rather enviable position of inheriting a sizable estate. In accordance with the will of his great-uncle, George Pitt, Second Baron Rivers (1751–1828), and by descent from his grandmother, who was sister of the second and daughter of the first Lord, he inherited Rushmore Estates. And with it came a small fortune—a gross annual income of a little under £20,000.[13] In today's terms this would represent nearly £2 million a year. One condition of the inheritance was that he change his name from Lane Fox to Pitt-Rivers—he compromised by using both of them.

The estate was a substantial one, covering about twenty-eight thousand acres, and located at Cranborne Chase, close to the Dorset-Wiltshire border. Formerly it had been a medieval hunting reserve. It took Pitt-Rivers very little time to realize that this new home was something very special. Wandering through this vast wooded area, perched high on chalk downland, he found it to be littered with numerous ancient burial sites and earthworks, and the remains of ancient settlements. And it was the excavations that he subsequently undertook here that were to form the fundamental basis of modern techniques of archaeological excavations.

Pitt-Rivers's introduction to field archaeology in England had been thirteen years earlier, from one of the great barrow diggers, William Greenwell, a canon from Durham Cathedral. Renowned as an "admirable raconteur, with a keen sense of humour,"[14] Greenwell was one of the many gentlemen in the early to mid-nineteenth century to whom dismembering barrows was an admirable hobby; to many it was little more than a pleasant Sunday afternoon pastime. Since the early eighteenth century it had been realized that barrows were the burial mounds of the ancient inhabitants of Britain, and excavating them became a fashionable exercise for country gentlemen. These diggings also were popular among those who had to work for their living, such as doctors, lawyers, or clerics. The results of the excavations were regularly reported in the columns of *The Gentlemen's Magazine* until late in the nineteenth century.

While to many barrow digging was little more than a treasure hunt or a way of impressing their acquaintances, to others it was a serious attempt to investigate the remains of ancient peoples in as scientific a way as possible. It often engendered bitter rivalry between some of its more serious practitioners. Like stamp collecting, some wanted to add more and more to their catalogue of barrows excavated, as well as increase the number of objects in their collections. In one of the nineteenth century's more forgettable poems, the Reverend S. Asaacson, writing in his *Barrow Digging by a Barrow-Knight* (1845), described the enthusiasm of the barrow digger,

> [whose] eyes upon the barrow bent are
> As if they'd pierce earth's very centre . . .
> Uprouse ye then, my barrow-digging men,
> it is our opening day!
> And all exclaimed, their grog whilst swigging,
> There's naught on earth like barrow-digging![15]

Greenwell was one of the more passionate excavators, driven by the desire not merely to add yet more objects to his collection, but seriously to record and better understand the distant past. Like his contemporaries, the standard excavation technique that he employed was to dig parallel trenches into the barrow, and then, in most cases, dig the entire barrow over. With this essentially destructive technique Greenwell excavated 443 barrows throughout Britain.[16] Besides producing a generally high quality of reports on his finds, Greenwell was an inspiration to the enthusiastic but inexperienced Pitt-Rivers.

Following a visit to some of Greenwell's excavations in Yorkshire, Pitt-Rivers began a survey of thirteen hill forts and flint mines on the South Downs in Sussex. Over the following twelve years he carried out the first detailed excavations using scientific principles into prehistoric sites in Britain. Unlike his contemporaries, instead of trenching directly into barrows, Pitt-Rivers very carefully dismembered the burial mounds layer by layer, inspired, no doubt, by William Smith's stratigraphic method. At the end of the work the mounds were reassembled. Pitt-Rivers's account of his work on these hill forts, read to the Society of Antiquaries in London in February 1868, was one of the first to highlight the significance of these constructions by the Neolithic to Iron Age inhabitants of the South Downs. "There is only a very scanty supply of vegetable mould upon these hills," he wrote in his report, "and the hard chalk rubble beneath the grass, undisturbed by cultivation until within the last few years, had preserved with wonderful distinction every break on its surface which during past ages has been caused by the hand of man, so that the faintest trace of an earthwork may sometimes be seen upon the green sward."[17]

At the same time that Worthington Smith was excavating two graves in Bedfordshire, General Augustus Henry Lane Fox Pitt-Rivers was engaged in two activities that occupied most of his time. One of his positions was as official inspector of ancient monuments under the Ancient Monuments Act of 1882.[18] The other was a project that he had been working on for five years: archaeological excavations on a much grander scale than Smith's. Not only were they substantially bigger, but they were taking him much longer to complete. Like Smith, Pitt-Rivers was also uncovering an uncommonly high number of fossil sea urchins. Unlike Smith, who at times lived a somewhat precarious existence from one illustration assignment to another, Pitt-Rivers, with endless time on his hands and the luxury of an almost endless supply of money, was able to employ more than a dozen helpers to assist him with his excavations. He also had no trouble getting the results into print, as he had the means to have them privately published. The result was four lavishly illustrated volumes, published between 1887 and 1898, and entitled *Excavations in Cranborne Chase, near Rushmore, on the Borders of Dorset and Wilts*.

While many of his contemporaries were still little more than treasure hunters, or were trying to prove the veracity of some dubious pet theory, Pitt-Rivers was a different breed altogether. His more than ten years of structured, well-organized, and meticulous work, which set out to improve on the known social and cultural development of our ancestors, was many years ahead of its time and set the standard for archaeological excavations of the twentieth century. He turned antiquarianism into archaeology by recording the exact position of his finds in precise detail.

The work on the Sussex hill forts in particular proved invaluable training for the excavations that Pitt-Rivers subsequently undertook at Cranborne Chase. Everything was recorded with meticulous detail, a product, no doubt, of his military training. Accurate measurements of every artifact and detailed plans on every aspect of the site were noted, from the stratigraphy to the skeletal remains (human and other animals), from every shard of pottery to details of the foundations of huts. Significantly, every scrap of rock or fossil that was unearthed was also kept and documented. Two of his basic principles of digging were that, first, "Superfluous precision may be regarded as a fault on the right side," and, second, "Tedious as it may appear to some to dwell on the discovery of odds and ends that have, no doubt, been thrown away by the owner as rubbish . . . yet it is by the study of such trivial details that Archaeology is mainly dependent for determining the date of earthworks."[19]

To help him in these excavations, Pitt-Rivers engaged a team of laborers and trained assistants, insisting that they be skilled in surveying and drafting. To begin with he employed eight to fifteen men, but later this number increased to twenty-five. They were paid fifteen shillings a week, supplemented by beer money. To supervise them he recruited a small team of young assistants and gave them various titles, such as assistants, subassistants, clerks, secretaries, subcustodians, and draftsmen. The supervisors slept at the museum that he had established in 1880 on his estate at Farnham. Their salaries were as high as £2 per week, with an additional £3 a month when they were actively excavating.[20] Pitt-Rivers's enthusiasm for archaeology was evidently not shared by all of his family, though. His wife "was quite indifferent to her husband's interests, while his daughters were quite antagonistic."[21] Writing in her diary in 1882, his daughter Agnes noted that "a whole heap of clerks came yesterday. Three—and the money the man [her father] gives them and the food they eat, if only we could have it we should be quite rich. Oh it is a wicked waste of money and they do no good. Drawing old stones and bones and shells etc."[22]

The men worked six days a week, and Pitt-Rivers visited the excavations three times a day to check on the progress of the diggings. (Another of his basic tenets: "No excavation ought to ever be permitted except under the immediate eye of a responsible and trustworthy superintendent.")[23] There is a view that, perhaps because of his meticulous nature, he was not the easiest of men to get along with. Although a powerful personality, he was cold and impersonal, inspiring respect rather than affection. But having drummed into his workers the importance of meticulous collecting and recording, his able assistants soon developed an eye for every small detail, dutifully collecting objects that others might have discarded or otherwise considered as of natural occurrence. It is probably for this reason that between October 4, 1886, and April 30, 1887, a period during which Worthington Smith had been finding hundreds of fossil sea urchins nestled around Maud and her child, Pitt-Rivers's team of workers was making a relatively sizable haul of urchins themselves.

Pitt-Rivers's excavation technique involved both "surface trenching" to a depth of a couple feet and excavating identifiable ditches. A little to the northwest of the village of Rushmore, he excavated the Romano-British village known as Rotherly. In the southeast quarter of the site alone, 32 fossil urchins were found, while 10 more were found in the east quarter. In all, 57 fossil urchins were uncovered. Given that this was chalk country, it wouldn't be particularly surprising to find a few of these fossils, as they would be reasonably common in this part of England. However, Pitt-Rivers was of the opinion that such concentrations far exceeded what would normally be expected from their natural accumulation. His view was that the people who had lived in this area some two thousand years earlier had gone out of their way to collect the fossils.[24] Like some collectors today, he thought that the urchins might have been used for barter, in effect acting as a sort of coinage. He suggested no other significance to their presence in the excavated material. But if they had been used as little more than objects of trade, why had some apparently been buried with human bodies?

Excavations in two pits that contained human remains also yielded fossil urchins. In one of these graves the skeletons of a newborn and a premature infant were found, along with a single urchin.[25] Another grave contained two adults and a newborn, and also included a fossil urchin along with other grave goods.[26] The question again is, were these chance occurrences? Not every pit that yielded a skeleton also contained a fossil urchin. But like Whitehawk and Dunstable, here were skeletons of newborns, found in graves with fossil urchins.

The settlement at Rotherly is much younger than the burial sites at Whitehawk and Dunstable. Coins found by Pitt-Rivers's team during their excavations suggest occupation of the site between the first and third centuries AD. Thus a little under two thousand years ago, as well as between four thousand and five thousand years ago, people were carrying out the strange practice of burying fossil sea urchins with their dead. The question that must be posed is, why were they doing it? And were they doing it for the same reason? Can there be a link, over the intervening thousands of years, in what these fossils might have meant to the three sets of people, people so far removed from one another in time, but perhaps connected in some small way through the memories contained in the fossil urchins? Or were people at different times quite independently attracted to these fossils, and coincidently consigned their dear departed into the afterlife with caches of fossil sea urchins? We have no written records of why these people collected fossil sea urchins, or why they sometimes buried them with their dead. However, there is one tenuous piece of evidence that just might provide an insight into these practices—the folklore associated with fossil sea urchins in more recent times.

Pitt-Rivers's collection forms the basis for the museum in Oxford that today bears his name—the Pitt-Rivers Museum at the University of Oxford. He had originally offered his fourteen thousand specimens to the nation, but had been turned down by the government. The nation's loss was Oxford's gain. It is a most eclectic collection, displayed in a manner that would horrify the modern "museologist" who has probably never collected or catalogued a specimen in their lives. In this ethnographic Aladdin's Cave, a Tahitian mourner's costume collected during Captain Cook's Second Voyage in 1773–74 rubs shoulders with a Hawaiian feather cloak, shimmering in shades of red and yellow. Cases of ceremonial brasses from Benin nestle against pre-Columbian pottery from the Americas, while a full-sized boat vies for space with a veritable orchestra of musical instruments. And tucked away in the dim, musty light at the far corner of the museum is a case of "magic" objects: amulets and charms to ward off evil in all its manifestations.

Two objects in particular caught my eye, when last my nose pressed up against the cold glass of the case: two flint fossil urchins, both the helmet-shaped *Echinocorys scutata*. What is so important about these two fossils is that not only were they collected because of their association with some aspect of folklore, but the meticulous collectors had also recorded with the objects the meaning attached to the fossils. To each is attached a small label. With the aid of a flashlight kindly supplied by an attendant, I was able to make out what was recorded on each. On one was written "'Shepherd's Crown' placed on window ledge outside to keep the Devil out. Sussex. Purchased 1911," and on the other, "Fairy loaf in the Eastern Counties. Its possession ensures bread in plenty. Bramford, Suffolk. D & H. Balfour 1919." Other fossils lying next to the fossil urchins were two polished heart-shaped pendants of fossil coral from Italy that were "charms against witches"; a toadstone (actually a fossil tooth of the fish *Sphaerodus gigas*) once used at Puddington, Devon, "to cure fits"; and two fossil belemnites, one from Oxfordshire which had had powder scraped from it "to give, mixed with water, to children suffering from eruptive diseases," the other from Dorset, "preserved as a thunderbolt used as a charm." A natural holed stone from Ballymena, Ireland, completed the collection: it had, in the 1890s, apparently been tied to a cow's tethering stake or between her horns "to prevent pixies from stealing milk."

While other fossils have also been attributed with a variety of powers, fossil sea urchins probably exceed all other fossils combined in the powers they were said to possess. They undoubtedly exceed all other fossils in the veritable thesaurus of names that have been bestowed on them—from "fairy loaves" to "snakes' eggs"; from "sheeps' hearts" to "shepherds' crowns," to give just a few. And it was Herbert Toms, one of Pitt-Rivers's assistants at Cranbourne Chase, whom we have to thank for preserving much of this folklore. He more than anyone else in England collected and collated most of the folklore associated with fossil sea urchins that we know today.

> He began to feel in a better humour, especially when the first beds of flint began to erupt from the dog's mercury and arum that carpeted the ground. Almost at once he picked up a test of Echinocorys scutata. It was badly worn away . . . a mere trace remained of one of the five sets of converging pinpricked lines that decorate the perfect shell.
>
> JOHN FOWLES, *The French Lieutenant's Woman* (1969)

8
Shepherds' Crowns and Fairy Loaves

The applause had died away; the vote of thanks had been given and gratefully received. Herbert Toms often gave versions of this talk, "Shepherds' Crowns in Archaeology and Folklore," during the 1920s and 1930s, while archaeologist at the Brighton Museum.[1] As usual the audience had been entranced by this neat man with trim beard, fiercely waxed handlebar moustache, bushy eyebrows, and flashing blue eyes. On this particular warm July day in 1929, he had ended his talk with a plea for anyone who knew anything about the folklore attached to fossil sea urchins to please come forward. This was one of the ways he gathered information about these tales from the past.

However, many of the details for his talk had been meticulously compiled by Toms while traveling the southern counties over the past three years. Accompanied by his notebook and camera, he had begun an organized campaign to record and preserve any folklore that still existed, talking with people he saw working in fields or in the little villages hidden deep in the chalk countryside of southern England. He knew that most of this knowledge was stored in the memories of generations who had lived in an era before airplanes and automobiles, and who had precious few years left to them. It was now or never. His was the first and, as it has transpired, the only attempt to systematically record the folklore associated with fossil sea urchins in England.

Despite his prolonged interest in this folklore, Toms (1874–1940) is better known for his excavations on the Iron Age hill fort of Hollingbury Castle Camp in Brighton. These he carried out in 1908, early in his career. He had also been involved in the excavation of a small part of a late Neolithic to early Bronze Age site in Sussex called Beltout. This site, located on the cliff top near Beachy Head, was excavated in 1867 by none other than Augustus Henry Lane Fox Pitt-Rivers. Toms was one of the last assistants to join Pitt-Rivers's staff at Cranborne Chase in January 1893, when he was twenty-two years old.[2] Pitt-Rivers was in declining health at that time, and indeed in 1892 had been seriously ill.

Toms left Pitt-Rivers a year before his final excavation in 1897. Having been inspired by the archaeological work at Cranborne Chase, he had applied for a position as an assistant at the Brighton Museum. His time with Pitt-Rivers must have counted strongly in his favor, because he was duly appointed to the position in March 1896.[3] It is also likely that Toms even owed his interest in folklore, albeit indirectly, to his mentor. Seven years after leaving his employment, Toms married a former personal attendant of Lady Pitt-Rivers. Born in Brittany, Christina Huon was deeply interested in the antiquities, customs, and folklore of her native land.[4] In later years she lectured extensively on these subjects, and this fascination probably rubbed off on Toms.

While he gleaned a lot of information about country folks' thoughts on fossil urchins from his excursions around the countryside, Toms's talks often furnished him with even more details. It was during question time on such a warm summer's day that he acquired some of his most useful information. An elderly man introduced himself to Toms as Mr. Turvey. He had lived around Brighton all his life and began to tell Toms of how he had worked on a farm outside town when he was young. He and his fellow workers were always coming across "shepherds' crowns" in the fields.[5]

For the next hour or so, as the long summer evening very slowly spread through the windows of the hall, Toms probably learned as much about shepherds' crowns as he had in the previous three years of wandering the countryside. Turvey had lived, it turned out, in the area for fifty years. Some of that time he had spent working on Upper Bevendean Farm, located only about one and a quarter miles (2 km) north of the Neolithic causewayed camp known as Whitehawk Camp. With his brother he had lived in a cottage on the farm and worked there during much of the 1870s.

Turvey explained to Toms that shepherds' crowns were found and picked up by virtually everybody who worked on Upper Bevendean Farm. Once they were in the hand they were either spat on and then thrown over the finder's left shoulder, or taken home and placed on a windowsill for luck.[6] This is in contrast to another of Toms's informants, from the Ouse Valley in Sussex, who specifically pointed out that only the right hand should be used for tossing the fossil over one's shoulder, and the fossil should *never* be spat on, since one would "never spit on your best friend."[7] Turvey himself would always select the finest shepherds' crowns and take them home for luck, placing them in a double row on a windowsill. According to him, such practices were locally very widespread because, as he told Toms, "something dreadful would happen if the shepherds' crowns were not so dealt with."[8]

The importance of treating the fossils in an appropriate manner whatever the circumstances was demonstrated by one of Turvey's workmates. "About 1875," Toms recorded after talking with Turvey,

> a Mr Blackman . . . was employed on Upper Bevendean Farm as a regular hand, and his principal work was hoeing crops. Blackman never passed a shepherds' crown, when hoeing, but picked it up, religiously spat on it, and then cast it over his left shoulder for luck. This led to Mr Turvey and his brother playing a practical joke on Blackman. Gathering a good number of shepherds' crowns, they strewed them between the rows of the crop that Blackman was due to hoe on the following day; with the result that Blackman, in conformity with his invariable practice, was perforce compelled to stoop and pick up each "crown" to spit upon and to throw over this shoulder. Much to the amusement of the Messrs Turvey, Blackman was subsequently heard to remark that he had never seen so many shepherds' crowns massed in one field.[9]

As Toms observed, whether his good luck was accordingly increased, there is no record.

Not all the shepherds' crowns were found when the fields were being plowed. Turvey and his brother were sometimes engaged in flint-picking—clearing the land of at least some of the countless flints that littered the surface of the ground. The flints were thrown into conical heaps, each occupying about a square yard (0.8 m^2). Then for luck, a shepherd's crown was placed on top of each mound. It is highly likely that this unusual behavior struck a chord with Toms, because he was very well aware of an article, written in 1907 by Paul Raymond, about another pile of stones found in a Bronze Age burial site in France.[10] The location, Mont Vaudois, near Héricourt in Haute-Saône in southeast France and near the border with Switzerland, is crowned by a small triangular plateau. It is protected on one side by a rocky cliff and on the other two sides by ramparts 433 yards (390 m) long. This crudely built wall would have been about 10 feet (3 m) high and about 50 feet (15 m) wide. There were all sorts of megalithic structures on both the inside and outside of this precinct: barrows, menhirs, and dolmens. The wall itself functioned as a barrow, housing graves roughly every yard. Within these were either cremated remains or remains in stone coffins.

Altogether, close to a thousand graves are scattered in this "sacred hill." In one barrow along the wall, a small tomb was found in the late nineteenth century. It contained a most strange mixture of objects: a human skull; a large, artificially worked bone; and a shallow cup made from red deer horn that had holes pierced in it, as though it had once been suspended by thread. But most surprising of all was that these items were covered by a pile of stones, like Turvey's mound of flints. But rather than being composed of irregular pieces of rock, the entire mound was built of an incredible 2.6 to 3.9 cubic yards (2–3 m^3) of fossil sea urchins. Toms calculated that this must have represented something like an astonishing twenty thousand to thirty thousand urchins.[11]

According to Turvey, many of the heaps of flints that had been piled up along the sides of fields were carted into Brighton to be used in making roads. Rather than use rounded stones, road workers whose sole job it was to crack the flints in half would split them open. However, whenever they came across a shepherd's crown that had escaped the eye of the likes of Turvey and Blackman, the stone-breakers would always carefully place it to one side, never to break it.[12]

Toms managed to glean other very useful tidbits of information from other attendees at his popular talks on shepherds' crowns. Two had told a very similar story. Laborers hired by what were described to Toms as people of "superior class" living in Billericay in Essex also used to keep their "fairy loaves" (as they were known in that region) on the mantelpiece, "to ward off evil spirits."[13] While this information was nothing new to Toms, what really surprised him was to learn that the fossils had to be religiously "blackleaded" once a week.[14] This was taken very seriously. If somebody forgot to do it, Toms was told, "the housewife would on remembering even get out of bed to descend and blacklead the crowns." He had been told of this practice also being carried out in Angmering in Sussex during the late nineteenth century. Another variant of this was the habit of varnishing the fossils before they were placed on the mantelpiece. Why either of these practices was carried out, Toms never discovered.

When he tramped the lanes and roads of southern England, ash walking stick in hand and pipe in mouth, much of Toms's interest in fossil sea urchins focused on discovering what people called them. In 1926 he published the only account of his research, which was brief, in *The Downland Post*, a short-lived newspaper produced in Sussex.[15] But in the fifteen years that followed he uncovered a veritable catalogue of names.[16] As Toms recorded in his *Downland Post* article, "Shepherds' Crowns are well known to nearly every ploughman, etc., one meets in [the] Downland [of southern England]."[17] Walking slowly behind their horse, these men of the soil would have had ample time to notice flints that were a little more regular than the usual gnarled stones that emerged from the red soil of the patches of "claywith-flint," through which they would often plow. Once they had spit on the fossil and rubbed off the soil, they could not have failed to be impressed by the five-pointed star that clung like a leech to the fossil surface. Along with individual flint molds of urchins, Toms reported on how occasionally larger pieces of flint can have more than one shepherd's crown set in them. One notable specimen (a flint nodule which he reported to be "in size midway between a respectable sausage and a saveloy")[18] contained five or six shepherds' crowns embedded within it.

Toms's early view was that there was little evidence to suggest that the name shepherd's crown was restricted to any one type of urchin. Others have suggested that the name has sometimes been applied to the heartshaped *Micraster* alone, rather than to both *Micraster* and *Echinocorys*. By the time of his last talk on the subject, though, in January 1940, he felt that the name was applied more often to the helmet-shaped *Echinocorys scutata*. Use of the name was widespread across southern England, with Toms recording its prevalence in Sussex, Dorset, Wiltshire, Hampshire, and Berkshire (and indeed it is still called by this name today in some areas, such as northern Hampshire). He posed the question of why these fossils were given such an unlikely name, but could never come up with a suitable answer.

Looking at either the heart-shaped *Micraster* or the helmet-shaped *Echinocorys*, it is not exactly obvious how the name was derived, given that one item of dress that shepherds usually did not wear was a crown. Whether we have to start invoking some ancient shepherds' ritual in which they donned strange urchin-shaped hats and indulged in silly dances is debatable. The only suggestions that have been made as to the origin of the name shepherd's crown are rather fanciful. Kenneth Oakley, in an article entitled

Herbert Toms, archaeologist and folklorist, Brighton Museum. Photo source: C. Duffin.

"Decorative and Symbolic Uses of Fossils," included an illustration of an eleventh- or twelfth-century carving from Autun Cathedral in France.[19] The depicted figure is wearing what appears to be a woolen cap; its conical shape has some vague resemblance to *Echinocorys*. This, Oakley considered, is what gave the name shepherd's crown to fossil urchins. Apparently in the Middle Ages until Tudor times, shepherds wore such a cap. This is said to be further supported by the possibility that shepherds wandering chalk downlands would find these objects. A major problem with this idea is that the areas where sheep gently graze are swards of lush grassland — hardly the most productive areas for yielding bucketloads of sea urchins. On the contrary, as Toms indicated, it would have been more likely that the fields being plowed would yield these paleontological riches. And why a crown? If there is one thing a fossil sea urchin does not look like, it is a crown. One very dubious explanation has been that the distinctive pattern of five radiating rays, most obvious in *Echinocorys*, somewhat resembles the supporting arms on crowns worn by the monarchy in medieval England. Or perhaps not. As I discuss below, the name shepherd's crown probably had nothing whatsoever to do with shepherds, or crowns, and has a far more ancient heritage than the Middle Ages.

Despite that, shepherds appear in other names. Around Ipswich in East Anglia Toms found that fossil specimens of the sea urchin *Echinocorys scutata* were called "shepherds' hats." In Brighton the heart urchin *Micraster* was also known as a "shepherd's heart." A particularly conical sea urchin from the chalk deposits, called *Conulus*, was given the name "shepherd's knee" in Rodmell in Sussex. Robert Plot, writing in 1677, recorded how this particular fossil urchin was sometimes called a "cap-stone," because of its similarity to a cap, with laces down its side.[20] Species of *Micraster* were also associated with other people's knees. Toms heard them called simply "knee-cap" in the Burgess Hill district, but "beggarman's knee-cap," "beggarman's knee," and "bishop's knee" in different parts of Sussex.[21] Bishops turn up again in "bishop's mitre," a name sometimes applied to the domeshaped *Echinocorys* urchins in some parts of Devon.

Some of the other names applied to fossil sea urchins in England were more localized in their usage. In northern Hampshire in 2001 I discovered that the heart-shaped *Micraster* is called a "sheep's heart."[22] The similarity to "shepherd's heart" suggests a possible derivation from this source. Here, only *Echinocorys* is known as a "shepherd's crown." Many of the common names allude to an association with fairies. Gideon Mantell, writing in 1844, recorded that they were sometimes known as either "fairy's nightcaps or turbans."[23] In the Isle of Wight they were once known as "fairy weights,"[24] the term *weight* probably deriving from *Wight*. In the time of Chaucer, the word *wight* was synonymous with elf, fairy, or other supernatural being, particularly for those that haunted houses. Another form of the fairy was the pixie. In mid-nineteenth-century Dorset they were known as "pexy" or "colepexy," and fossil urchins were called "colepexies' heads."[25]

Fairies also appear in a name that often crops up, particularly in Essex and especially in relation to *Echinocorys*, and that is "fairy loaf" (more rarely, "fairy head"). This name Toms recorded as having been used in Dorset in the 1870s.[26] Oakley received a letter in the 1940s indicating that the name fairy loaf was used to describe flint urchins found in the gravels in the London area, Berkshire, and Surrey at that time.[27] In Suffolk the fairy loaf was sometimes known as the "pharisee loaf," the term *pharisee* deriving from *fairisee*. This sometimes became "Paris loaf." At other times "pharisee loaf" was corrupted to "farcy loaf." One suggestion for why farm horsemen kept these fossils in their pockets was that they were thought to act as a charm against an infectious disease called glanders in horses. In his fascinating account of the folklore of East Anglia in England,[28] George Evans argued that *pharisee* derived from *ferrisheen*, which is a derivation of the Gaelic *fer-sidhe* (pronounced "far-sheen"), meaning "man of the hill," including every god and every fairy.[29] As I will argue, here is the link with "shepherds' crowns," through the *sidhe*.

In a letter written to the Sunday *Times* published on Christmas Day 1955, one Beatrice Lubbock recalled being handed a "Pharisees' loaf" in a public house in an Essex village, and being told that anyone finding such a stone would never want for food.[30] In Suffolk the urchins were placed on the mantelpiece not only for decorative purposes, but because it was believed that they would ensure there would always be bread in the house. In northeast Suffolk the urchin was used as a charm when placed by the brick oven in the bakery. It was thought that the loaflike shape of the urchin *Echinocorys* would guarantee that the bread would rise.[31] This preoccupation with food was also extended to calling them "sugar loaves" in Kent and Sussex, according to Gideon Mantell in his 1844 book *The Medals of Creation*.[32] Bread and sugar were combined in north Norfolk, where the fossil urchins *Echinocorys* and *Micraster* were called "fairy sugar loaves."[33]

One could argue that the name fairy loaf seems to allude to the idea that fossil urchins may once have been thought of as little loaves of bread that belonged to fairies. But can such folklore really be traced back to concepts that Maud and her people must have had for these objects? At some stage in the past, were the links so fractured that the original meaning of these objects to ancient people became so distorted? Would Maud really have wanted to be buried with enough tiny loaves to feed an army of fairies? To the Celts, fairies were not the sweet little flying folk of modern-day fairy stories. They were creatures to be very wary of, even feared. Many of the myths in Britain associated with fairies, elves, pixies, goblins, and so on probably have their roots deep in pre-Celtic times. It is interesting to note that another word for these creatures was *urchin*. This word may have derived from *orcneas*, a word used in the epic Anglo-Saxon poem *Beowulf*.[34] In more recent times in J. R. R. Tolkien's *The Lord of the Rings* it has been transformed into "orc." It is interesting to speculate on whether the use of the word *urchin* for the marine animal (technically an "echinoid"—which means "spiny creature") originated in its association with "fairy" folk, the name therefore first being used for the fossil. When it was realized that these strange objects were the fossilized remains of the marine invertebrate, the name was transferred to the living variety.

To the Celts, fairies were the inhabitants of a mystical, enchanted world—the Otherworld. This place was described to the warrior Oisin by the fairy-woman Niamh of the Golden Hair as "the most delightful land of all that are under the sun; the trees are stooping down with fruit and with leaves and with blossom. Honey and wine are plentiful there; no wasting will come upon you with the wasting away of time; you will never see death or lessening. You will get feasts, playing and drinking; you will get sweet music on the strings; you will get silver and gold and many jewels."[35] While this magical world could be entered through caves and lakes, it was most accessible through prehistoric burial mounds, the barrows of Neolithic and Bronze Age times. The people who inhabited these mounds, the fairies and pixies of more recent times, were known as the *sidhe* (pronounced "shee"). In Anglo-Saxon the barrow was called *biorh* (pronounced "berr"). Here lived the *sidhe* folk, gods and goddesses of the Otherworld. Over time they became the fairies of folk belief. Living in the Otherworld the inhabitants never grew old, nor did they suffer pain or sickness. Each *biorh* had a magic cauldron that produced an endless supply of food. This, and fruit from magic trees, had the power of restoring the dead to life.

The Otherworld was also the land of the dead. Its inhabitants were sometimes considered the spirits of the dead that were still trapped between this world and the next, awaiting their chance to return to mortal life. Given the frequency with which fossil urchins turn up in barrows and burial sites from the Neolithic times onward, and the number of folk names for these fossils that had a relationship to fairies, it seems likely that fossil urchins played a role in these myths associated with the Otherworld. The long tradition, dating back to Neolithic times, of placing fossil echinoids with the dead in graves and barrows, or even placing them in barrows that contained no bodies, suggests that they were in some way associated with helping the spirits of the dead on their journey into the Otherworld. Fairy loaves were therefore, perhaps, spiritual food to sustain the spirits into the Otherworld and to ensure their immortality.

But what of "shepherds' crowns"? I suspect that this name has nothing at all to do with shepherds, but derives, in part, from the Celtic word *sidhe*. The barrow within which the *sidhe* lived was the *biorh*. It does not seem too unlikely for *sidhe biorh* ("shee berr"), the fairy home, to be transformed over thousands of years into "she-pherd." The close similarity in the shape of the domed *Echinocorys* to a barrow suggests the possibility that the little fossils may have been seen as emblematic of a barrow. Why "crown"? I do not know.

A point that I will return to in more detail in later chapters is how, in more recent times, fossil urchins have been considered to be lucky charms.

Toms recorded how in Dorset in the 1920s people regarded the shepherds' crowns that they found as lucky objects, and would carry them home and place them on the windowsills or by the front doors of their cottages.[36] This was for the same reason that prompts people to put horseshoes and stones with holes in them near the door—for luck, and also in the past as a charm against witchcraft. More than that, shepherds' crowns were specifically thought to be a protection against the devil. In Norfolk the fairy sugar loaves were placed in rows on windowsills to bring luck to the inhabitants of the cottages.[37]

Using fossil sea urchins apotropaically has an extremely ancient heritage, as shown by the fact that nearly two thousand years after people in Roman times picked up fossil urchins in the Cranborne Chase district and placed them with their dead, locals, as Toms discovered in the 1920s, were still collecting them in the same area. They called them "lucky stones." Toms learned that in south Wiltshire, when the locals found such a lucky stone they wouldn't necessarily always take it home with them. Like the people working the fields around Brighton, they would spit on it, and then throw it over their left shoulder for luck. Whether these stones were ever retrieved and taken home to put on the windowsill, or simply left there for the next passerby to spit on, Toms doesn't record.

Herbert Toms spent much of the summer of 1929 undertaking a photographic survey of "shepherds' crowns" that had been placed on cottage windowsills. This was no spur-of-the-moment whim. The year before, he had visited a number of villages in Sussex and Dorset and realized that this was a commonplace activity. Toms concentrated his efforts in two areas: the region around Worthing in Sussex, and around the villages of Handley and Ashmore in northern Dorset. In Sussex he photographed seven cottages where the windowsills were decorated with shepherds' crowns, and six in Dorset. The number of fossils on a single windowsill varied enormously, between 4 and 69. By far the most common fossil that Toms found, perching like rows of buns on a baker's tray, were the helmet-shaped urchin *Echinocorys*.

Toms was intrigued to find out what drove the inhabitants of the cottages first to collect the fossils, and second to put them on their windowsills. The answers were not always the same. What he found was that people in Sussex had quite a different rationale from their contemporaries in Dorset for this strange behavior. Most people he asked in Sussex regarded the fossils merely as "curiosities." However, in two cases where the fossils were known to have been collected fifty years previously, they had been kept because they were considered as "lucky" objects when they were

Fossil urchins (all *Echinocorys scutata*) photographed by Herbert Toms in 1928 on the windowsill of a cottage in Patching, Sussex. According to Toms's notes: "Mr Ruff found the crowns in position when he took possession of the cottage in 1923. They were placed on the sill by a former occupant . . . about 1909. Regarded . . . as curiosities; but the custom is evidently a survival from the time (about 50 years ago) when nearly every local cottage sill had its full complement of 'crowns' which were then regarded as lucky." Reproduced with the kind permission of The Royal Pavilion and Museums, Brighton and Hove.

found. So, in just that short space of time the "old" ideas had drifted imperceptibly away, like a fading early morning mist.

While belief in fossil urchins as lucky objects had essentially died out in Sussex by the 1920s, it lingered on into the late 1920s in Dorset. Toms found that five of those who placed shepherds' crowns on windowsills in Dorset at that time did so because the fossils were thought to bring luck. According to one Mr. Wort, the "crowns should be held base up, spat on, thrown into the air three times and caught, then placed in (the) pocket for luck."[38] This behavior of throwing the fossils into the air is resonant of an ancient practice of the Celts reported by the Roman natural historian Pliny two thousand years earlier (see chapter 12). It would seem that by the first half of the twentieth century in Sussex, fossil urchins had transcended Kenneth Oakley's three stages in the evolution of the cultural perception of natural objects, moving beyond the "lucky" stage to a fourth stage, where they had simply become objects of idle curiosity.

Archaeological evidence suggests that the practice of placing urchins near the windows or doors of houses is not just a recent activity but extends far back in time, to at least two thousand years ago. Excavations at a Romano-British settlement at Studland in Dorset record the prolonged occupation of this site for at least four centuries.[39] A detailed archaeological excavation carried out in the early 1950s in an exemplary stratigraphic manner (of which William Smith would have been proud) revealed how over hundreds of years new buildings were built over older ones, layer upon layer. As old huts or cottages fell into disrepair, they were demolished and replaced by new ones. A stratigraphic sequence comprising six dwellings was recorded, the oldest dating from the mid-first century AD, through to the fourth century AD. In each of the dwellings, fossil urchins were found. The location of some of these fossils suggests that their presence had some special significance to the occupants, perhaps as some sort of "house charm," or charm against the devil. This, as Adrienne Mayor pointed out to me, is an excellent example of folklore illuminating archaeology, and archaeology confirming folklore.[40]

The earliest dwelling was a round hut, fourteen feet in diameter, with an inner ring of roof supports. The only urchin from this hut came from soil under a stone wedged against one of the inner posts, suggesting that it had been deliberately buried there. The hut had been replaced by a rectangular cottage between AD 60 and AD 85. Two urchins were found in the earthen floor of this dwelling, where they had possibly been buried. Only inches apart, they lay directly over the fossil from the earlier hut. It is possible that the occupants of this later hut remembered the location of the other fossil, or the information had been passed down to the next generation, along with the significance of this placement. Fossil urchins from later dwellings built on top of the remains of the earlier ones were less deliberately placed. The second-century cottage contained three fossil urchins (along with another found outside the cottage). Three cottages existed between the third to fourth centuries. The earliest of the three contained a single urchin, the next had two urchins, and the youngest, a single urchin.

Given this close association between fossil and location, it is likely that throughout this period some sort of ritualistic significance was attached to the fossils. Interestingly, all fossils urchins from the four-hundred-year period were the helmet-shaped *Echinocorys*. In many ways this urchin also resembles a round hut in shape, much like the earliest hut at the site. Could the occupants of the dwelling have believed that the strength of the stone might in some way have been sympathetically transferred to the hut, keeping it safe from the powers of evil? Whatever the reason, the echoes of this activity seem to have persisted for almost two millennia. Were they shepherds' crowns or fairy loaves to these people? Or were they called something quite different? Unfortunately, we have no way of knowing.

There is, though, another, quite different name that has also been given to fossil urchins in southern England—a name that may have been in common usage for thousands of years. While researching the folklore of fossil urchins, Herbert Toms uncovered evidence to suggest that up until the mid-nineteenth century this name was popularly used in many parts of southern England. It is a name that formed part of a folk tradition that was most probably brought to Britain by Danish and Anglo-Saxon invaders more than fifteen hundred years ago. And it is a name that has its roots deep in Norse mythology—"thunderstone."

Fear no more the lightning-flash,
Nor the all-dreaded thunder-stone;
Fear not slander, censure rash;
Thou hast finished joy and moan:
All lovers young, all lovers must
Consign to thee, and come to dust.

SHAKESPEARE, *Cymbeline*, act 4, scene 2

 9

Thunderstones

It was 1910, and the two workmen, Mercer and Crittenden, had been digging for about an hour in silence, setting a steady rhythm: in with the spade, lift the dark clay, toss it on the heap at the side, then thrust it back into the soil; in with the spade, lift the soil . . . They were digging the foundations for a large house situated in Southborough, on the northern outskirts of Tunbridge Wells, in Kent. The house, which was to be called Hillgarth, was located at the corner of Powder Mill Lane and the London Road, and was one of a number of imposing homes being built in this part of town at the end of the first decade of the twentieth century.

There is no record of whether it was Mercer or Crittenden who made the discovery, but when they had dug down to about two and half feet (0.75 m) below the surface one of the spades suddenly struck a hard object. They stopped digging. Maybe for one glorious moment they had visions of a box, and of themselves prying its lid open to find masses of gold coins. More prosaically, what probably happened was that one of them bent down into the hole and carefully began to brush the soil away from the object that his spade had struck. It was probably just another bit of old brick. Whatever it was, though, it wasn't very big. He would have been able to get his hand under it quite easily.

Carefully, the muddy object was lifted out of the trench. What lay in Mercer's or Crittenden's dirty hand was a small and rather unprepossessing grayish-black pot. Its surface had a rather gritty texture, and was about four inches (10 cm) high and three and half inches (9 cm) in diameter. The outside of the pot was quite smooth, lacking any sort of ornamentation. It is unlikely to have struck either of the workmen as a particularly exciting discovery. Hardly a pot of gold. What would they have done with it? Tip it upside down, perhaps—give it a sharp shake to empty the mud out. But more than sodden soil tumbled out. Not gold coins but, quite unexpectedly, two most remarkable objects: a broken, polished flint axe, and a perfect specimen of the fossil sea urchin *Micraster*.

If this was all that Mercer and Crittenden thought their diggings would turn up that day, they were very much mistaken. Some time later, after they had resumed their digging, a shovel again struck a hard object. *What now?* they probably thought. It would have taken longer to scrape the mud away from what turned out to be an even larger vessel. It stood upright in the ground, close to where they had found the smaller one. It was much the same color but taller, being "at least 15 inches and had a cover," Mercer and Crittenden had later reported to George Abbott, director of the Tunbridge Wells Museum. After hauling the pot out of the trench, they carefully pried the lid off. One can imagine them peering inside, wondering what lurked within, perhaps hoping against hope that they had finally found their pot of gold. If that's what they thought, they were again disappointed. What they saw probably took them aback nonetheless, for tumbled together inside the urn was a mass of blackened, burned bones—and the bones were human.

What the two workmen had stumbled on appeared to have been the site of a cremation burial—an urn with burned human bones and a small pot with some rather odd grave goods. In all likelihood, given the similarity in composition of the containers and their close proximity to each other, they were almost certainly buried at the same time. However, despite the men's continuing to dig the rest of the foundations over the next few days, no more pots of bones or fossils were found.

While the owner of the house insisted on retaining the large urn and its ashy contents, he wasn't interested in the other pot and its strange goods, so Mercer was allowed to keep those. He duly presented them to Abbott, who in turn incorporated them into the museum's collection, where they remain to this day. As the card accompanying the specimens records,

Iron Age funerary urn and its contents: two thunderstones, an incomplete Neolithic axe, and a fossil sea urchin, *Micraster*; from Southborough, Kent. Specimen in Tunbridge Wells Museum. Author photo.

Small Pot, 3.5" × 4" greyish black coarse paste, softish, found in August 1910 by workmen Mercer & Crittenden 2'6" below the sirface (sic.) while digging foundations of "Hillgarth" at corner of Powdermill Lane & London road. It held the broken half of a polished flint Axe & a fossil Sea echinoid (*Micraster*). Three ft away was a large Urn with burnt bones in it, it was "at least 15" wide & had a cover" (I saw Mercer in 1930 who confirmed this account J.C.M.G.) The large Urn was retained by the owner of the house & afterwards broke & lost (Tattershall Dod).

As the card recounts, unfortunately the urn never made it to the museum, having been subsequently broken and, along with its human remains, lost. The small pot was later studied by Reginald Smith, archaeologist in the Department of Ethnology and British Medieval Antiquities at the British Museum. He identified it as a type of pottery typical of the period known as "La Tène 1."

The La Tène culture developed in central and northwestern Europe during part of the Iron Age, from about 450 BC to approximately 58 BC. The name La Tène (meaning "The Shallows") derives from a Celtic archaeological site in Switzerland where spectacular discoveries were made of iron weapons, implements, and jewelry. Many of these items were decorated with a particular design, characterized by a predominantly abstract style of curving lines. The La Tène style of artwork reached Britain by about 300 BC, and by 200 BC it had spread to Ireland. So if the rather simple pot that Mercer found is indeed of this period, it points to its having been made sometime after 300 BC—in other words, during the earlier part of the Iron Age.

Earlier examples of cremations are known in Denmark. During the period between 500 BC and 150 BC, the standard method of burying the dead in southern Jutland was by cremation. The ashes were sometimes placed in miniature round mounds. In some later burials, between 150 BC and 50 BC, the cremation graves contained, among a variety of grave goods, black-burnished pots. Why they were blackened is not clear. Could there, perhaps, be an echo here of Herbert Toms's records of similar blackening being applied to fossil urchins two thousand years later in southern England? None of these cremation graves in southern Jutland are recorded as having contained fossil urchins. They contained far more prosaic items: cremated animals, weaponry, and jewelry along with other more elaborate goods, such as bronze vessels and even wheeled vehicles. Despite this lack of urchins, the contents of the Tunbridge Wells pot provide another fascinating link with Denmark.

The practice of cremating rather than burying the dead seems to have begun in Britain, as in other parts of Europe, about 3,300 years ago, in the late Bronze Age.[1] Yet despite this change in burial practices, there is evidence from a windswept hill on the Isle of Wight in southern England that the habit of burying fossil sea urchins with the dead had not died. The Isle of Wight is a lozenge-shaped island about twenty-three miles (37 km) wide. Running across it from east to west is a twisted ridge of urchin-bearing chalk. Thrust upward into a steep fold during the last gasps of the tectonic activity that threw up the Alps, this ridge courses across the island like a deformed spine. And scattered along its highest parts, the ridge is silhouetted by circular burial mounds: green barrows meeting blue sky, closer to the Sun, closer to the stars. Often they sit there in silent isolation. But in a few places they form clusters.

One of the densest concentrations occurs on a high part of the chalk ridge called Ashey Down, where, over a 777-yard (700-m) section, there are twelve barrows. The largest is about 13 feet (4 m) high and about 82 feet

(25 m) in diameter; the smallest is just under 3 feet (1 m) high and no more than about 50 feet (15 m) wide. But the barrows are not on the highest part of the ridge. Six are scattered on the western flanks of the Down, while the six others form a tight circle on the eastern flank. Standing on the largest barrow it's not hard to see why the burial site on the flanks of the hill was chosen, for the views to the west, and in particular to the east, are simply spectacular. From the eastern circle of six the sea can be seen to the north, east, south, and west. When first constructed they would have been visible from afar as stark, white mounds of chalk. Now they are draped in green, and on early spring mornings dappled with white daisies and yellow cowslips.

In the mid-nineteenth century, when amateur antiquarians became interested in excavating such barrows, a local physician (most appropriately named Benjamin Barrow) decided to find out what was lurking in the mounds on Ashey Down. Where to start digging? The obvious first target was the largest of the group of six, on the eastern flank of the Down. Once Barrow had started digging it didn't take him long to find something. But if he was hoping for a cache of grave goods he must have been disappointed, for all he found were burned bones and charcoal—lots of charcoal. Here must have been a huge tangle of branches and logs on which had been placed the body. Once the pyre had been lit, what a spectacular sight it must have been, visible for miles around, even from the sea. The sky consumed the fire and "Heaven swallowed the smoke,"[2] as the author of the great Anglo-Saxon poem *Beowulf* relates after the dragon-slaying hero has himself been slain, and his remains placed in

> a mound on a headland, high and imposing,
> a marker that sailors could see from afar.[3]

Unlike the golden torques and jewels placed in Beowulf's barrow some fifteen hundred years later,[4] the grave goods that accompanied the deceased on Ashey Down were meager fare: "portions of pottery with rude indentations, a few small pieces of iron pyrite, and an echinite [a fossil sea urchin]."[5] Barrow, the excavator, might have been surprised at what he found, but in his account he offered no explanation for why he thought a fossil had been placed with the body.

He might have hoped for something a little more interesting from the next barrow he excavated, but again he was to be disappointed. All he found were more incinerated bones. But the third of the six graves was more interesting: burned bones again, but this time they had been placed in an urn that was surrounded by a mass of carefully placed flints. And the sole grave good was another fossil sea urchin. The fourth grave yielded nothing but charcoal, the bones probably having disintegrated. But it was in the fifth barrow he excavated that Barrow found something that he was probably looking for. He had to dig a little deeper this time. But what he discovered certainly rewarded his efforts: a bronze dagger. There were yet more burned bones and charcoal, and among the blackened soil yet another single fossil urchin that had been placed with the body.

The last of the six barrows yielded nothing but charcoal, as did five of the remaining six barrows on the western flank of Ashey Down. But in excavating the last one Barrow's perseverance paid off, as a small urn "7½ inches high and 2 feet 1 inch in circumference"[6] and filled with incinerated bones emerged from the charcoal and chalk, along with an animal tusk, a piece of iron pyrite, and yet another fossil urchin. So of the five barrows that yielded human remains, four also yielded fossil urchins. It is hard to know whether the fossils would have been placed with the bodies before or after they were cremated. But if the events related at the end of *Beowulf* reflect traditional cremation practices, then they were probably placed on the burned bones before the incinerated pyre was covered with earth and stones:

> It was their hero's memorial; what remained from the fire
> they housed inside it, behind a wall
> as worthy of him as their workmanship could make it.
> And they buried torques in the barrow, and jewels[7]

—and sometimes fossil sea urchins.

Daggers or hand axes and fossil sea urchins might seem rather strange bedfellows to accompany someone's cremated remains on their journey into the afterlife. Yet such as association, as I have indicated earlier, is a recurring theme from Neolithic through to Roman times. Arguments for fossil sea urchins having a similar ritual significance in earlier Neolithic times are supported by the discovery at la Motte St. Jean, Saône-et-Loire in France of a polished axe and three fossil sea urchins.[8] Likewise, late first- to third-century remains of Gallo-Roman dwellings, temples, and wells at the Forêt de Rouvray, near Rouen in France, contained a cache of twenty small Neolithic axes, along with twenty-two fossil sea urchins.[9]

Yet one man living in Denmark at the time that Mercer and Crittenden discovered a Neolithic axe and a fossil urchin buried together in Tunbridge Wells wouldn't have been at all surprised by it. In 1910 this man, Christian

Blinkenberg, was feeling rather pleased with himself. The previous year his long article entitled "The Thunderweapon in Religion and Folklore: A Study in Comparative Archaeology" had been published in Copenhagen.[10] Moreover, he had recently obtained a grant from the Carlsberg Fund to enable the article to be translated into English. No less a publisher in England than the University of Cambridge Press had agreed to produce it as a book the following year.

Blinkenberg (1863–1948) had, for some time, been fascinated by the historical relationships between ideas about the "thunderweapon" that had existed in the Greek region at various periods in the past. In the Mycenean age, for example, the thunderweapon was a double axe of bronze; in early historical times it was known as the *keraunos*, or the thunderweapon of Zeus. In more "modern" times it was sometimes the name that people gave to ancient stone axes that they uncovered from the ground.

Born in Ribe, Denmark, in 1863, Blinkenberg first worked at the University Library in Copenhagen, but following a study trip to Greece and Italy between 1889 and 1891 and the awarding of a doctorate for his study of "Asklepios and his men in Hieron at Epiddaurus," he joined the National Museum of Denmark as an archaeologist. During the first few years of the twentieth century he led excavations in Lindos in Rhodes. Yet despite his studies in Greece, he found it difficult to understand the origins of the concepts of the thunderweapon by exploring the Greek region alone. He realized that he had to look further afield. And where better to start than on his own doorstep?

For some time he had been aware that certain objects in Denmark were known by country people as "thunderstones." However, like Herbert Toms, he was challenged by the lack of any kind of systematic study of the popular traditions relating to thunderstones to help him understand the source of this concept. Clearly, the only answer was to carry out the study himself. Whether Blinkenberg was merely a less energetic character than Toms, or whether he just thought that there were better, more efficient ways of obtaining the necessary information for a study, I don't know. But Blinkenberg had resolved that rather than traipsing around every village and farm in rural Denmark and interviewing people over farmyard gates, he would make judicious use of the media. He would simply gather all the information that he required without setting foot outside his office. He resolved to place a small advertisement in a number of newspapers, appealing for anyone with knowledge of the folk traditions attached to any objects that they called thunderstones to write to him via the Danish Folklore Collection. Then he would collate the information.

Blinkenberg would have had no idea what kind of response he would get. So he was probably delighted when, between 1908 and 1911, he received more than seventy replies, all containing a wealth of fascinating information.[11] About half the letters were from people living in northern Jutland and on the islands of Zealand, Funen, and Langeland, to whom ancient flint axes were thunderstones. A few respondents were people living in the southern islands of Falster, Lolland, and Bornholm. These people called fossil belemnites thunderstones.[12]

The rest of the replies that Blinkenberg received about thunderstones were from Jutland—to people in parts of this region, fossil sea urchins were known as thunderstones. However, irrespective of whether their thunderstone was a stone axe, a belemnite, or a fossil urchin, the associated stories they had to tell were much the same. In essence, all of Blinkenberg's respondents had been brought up to believe that thunderstones fell from the sky during thunderstorms, specifically when lightning struck. As a result, the thunderstone was said to possess special powers that protected the finder against lightning.

A typical story that echoed the ideas of many of Blinkenberg's respondents came from Andre Jensen, the headmaster of Flemming School:

> Thunderstones were fossilized sea-urchins. I never heard other stones called thunderstones. At the strike of lightning such a stone, in a glowing state, fell down and brought the fire with it. Only when a crashing thunderclap followed the lightning did we think that a stone had fallen, and it was precisely its fall and great speed which produced the crashing sound. In all other cases the stone remained in the thundercloud. Such a stone acted as a protection against lightning (the thunder would not strike where it was). The stones were therefore collected and carried home. They were put everywhere as safeguards, both in the house and the out-houses, in a window or on a shelf, on a beam or in a corner, but they were not made much of and in most cases were covered by dust and cobwebs. Notably large and fine specimens were laid on a chest as a decoration. One often carried a small stone about, as a protection, when out in a storm. I myself carried a couple of stones in my pockets and felt fairly safe in a thunderstorm; I did this even after I had learned at school what the stones really were.[13]

Sometimes, rather like the fossil urchins found in the archaeological excavations at Studland in Dorset, the thunderstones were immured in a wall or laid under the floor. At other times they were placed high up, such as atop a four-poster bed, or even on the roof. Mr. Clemens Sönnichsen of Ballum wrote the following to Blinkenberg:

> The thunderstones (fossilized sea urchins) were believed to avert lightning; they were put on top of the clock and in various places about the house—even in the loft under the roof. The house was shielded and we felt fairly safe and well protected with them near us while the storm lasted. When these stones were damp (when they "sweated" as we said) it was always a sure sign that a storm was coming on, and as long as the stone "sweated," we children were if possible kept at home.[14]

Thunderstones were also attributed with other powers. They were thought to keep trolls and other nasty creatures, such as witches, from the house, as well as keep away evil in general and bring luck to the house.[15] More specifically it was said that they guarded the unchristened child against being "changed," as well as protected horses in their stables from having nightmares.[16] Their ability to keep witchcraft out of the dairy was another major attribute, as witches "could take the butter-luck away from people."[17] So, like shepherds' crowns, thunderstones were placed in the dairy to keep the milk fresh and to give better cream, and they were put on the churn to produce better butter. Placing them on window ledges or over doors in stables was thought to protect the cattle against disease and accidents.

While thunderstones was a popular name for fossil sea urchins in much of southern and central Jutland,[18] Blinkenberg recorded a host of other local names that were sometimes used in Denmark. These included *sebedai, Sebadeje, spadejesten, pariko, paradisko, paddeko, palliköer, marmorsko, marrimusko, smördie, smörsten,* and *smörlykke*.[19] He provided no explanation for these terms, but a more recent study by Viggo Sørensen has clarified their meaning, as well as their regional distribution.[20] *Spadejesten / spådejesten, sebedæjsten,* and *spadæje,* used in central and eastern Jutland, can be translated as "Zebedee stone."[21] Zebediah was the father of the apostles James and John, who were sometimes known as the "sons of thunder." Therefore Zebedee stone can be thought of as "thunderstone."[22] (Mark 3:17, King James Version: "And James the son of Zebedee, and John the brother of James; and he surnamed them Boanerges, which is, The sons of thunder.")

The terms *paradisko, paddeko, palliköer,* and *pariko,* confined to northern Jutland, essentially mean "paradise cow." Sørensen suggests that *paradise* may refer to where the fossil urchins must have come from, owing to belief that they fall from the sky during thunder. Cow refers to their frequent use in children's games.[23] Cows feature also in *marmorsko* and *marrimusko*, terms known from a small area of northern Jutland, which translate to "marble cow," marble supposedly referring to the nature of the fossil's surface.[24] The last group of names, *smördie*, *smörsten*, and *smörlykke*, used mainly on the islands of Zealand and southern Funen, refer to the attributed powers of the thunderstones to prevent butter or milk from going bad.[25] On the southern, smaller islands of Lolland, Falster, and Mons they are simply called *stejerne*, meaning "star," whereas on the island of Als, situated off the southeastern part of Jutland, they are called *kaemerknap*, meaning "big button." *Kaemerknap* is also used on a small part of the mainland adjacent to Als.

It can hardly be a coincidence, then, that the two objects found in the pot in Tunbridge Wells were, even as late as the early twentieth century, still both referred to as thunderstones. But calling fossil urchins and old stone axes by that name is one thing; to place them together in a grave is quite another. As the sole objects accompanying the Tunbridge Wells burial it would seem that during the Iron Age they held greater significance than in more recent times. While a similar association has not been found in ancient burials in Denmark, fossil urchins have certainly been discovered on their own in some Danish archaeological sites. Blinkenberg recorded that they were frequently located in positions in excavations that could not have been accidental.[26]

One placement in particular has a striking resonance with one that was being carried out near Brighton in the late nineteenth century. In the year that Blinkenberg's book was first published, a barrow near Sönder Omme in Vejle district was excavated. One of the more surprising finds was that of a fossil sea urchin placed over the middle of this sunken Bronze Age grave. It was found "in a position which suggested that it had originally been placed on the heap of stones that covered the grave."[27] Could Brighton native Mr. Turvey's habit of placing a lone fossil urchin on top of the mound of stones that he had cleared from the field (see chapter 8) have been a distant memory of a more solemn, spiritual ceremony?

Although the Iron Age deposits in Denmark may lack evidence of the burial of thunderstones, these were almost certainly being kept with the living in ancient settlements. In 1907 two fossil urchins were found close to each other in the remains of an Iron Age hut. Like those at Studland, they may also have been placed in the house as thunderstones to protect it against lightning. Blinkenberg reported that the urchins "lay close together in the actual deposits containing the remains of the house which had been destroyed by fire. Thus they would seem to have been placed in the house or possibly under the roof."[28]

Those Danes and other early immigrants from northern Europe who crossed the North Sea in search of a better world on the faraway island of Britain must surely have taken their gods and their beliefs with them. When Herbert Toms was uncovering the folklore attached to fossil urchins in southern England, twenty years after Blinkenberg had written his book, he heard many stories that would have not been out of place at all in the Dane's work—fossilized memories handed down from generation to generation.

Toms learned from the well-known folklorist Edward Lovett that between 1865 and 1870, Lovett had seen fossil urchins on the outside window ledges of a number of cottages in both Wiltshire and Gloucestershire.[29] Toms had been told that these fossils had been placed there to protect the dwellings from thunderbolts. He discovered that in the 1920s and 1930s, fossil urchins in parts of south Dorset were specifically known as thunderbolts, as they sometimes were in parts of Northamptonshire, Sussex, and Hampshire.[30] In some places in Sussex, however, they were also called thunderstones. In all these places, as in Denmark, they were said to be either carried around in the pocket or placed on windowsills or mantelpieces, to protect the person or house from lightning.

From his research, Toms concluded that the use of the terms *thunderstone* or *thunderbolt* in relation to fossil sea urchins had been more widespread than *shepherd's crown*. The use of these stones as lucky amulets was undoubtedly very deeply rooted in the psyche of these people of southern England and the northern parts of Europe. As Blinkenberg remarked,

> It is probable that a popular belief which is so widely spread is a very ancient one: and there can hardly be any doubt that the idea of the thunderstone was spread from people to people over the many lands in which it appears, and did not arise spontaneously in different countries in such consistent forms and with so many features in common.[31]

To Scandinavians he was Thor; to the English he was Thunor;[32] to Celts he was Tanarus; and to Germans, Donar. This god of thunder and storms and wielder of mighty thunderbolts also shares some attributes with the Roman god Jupiter. His ancient lineage can be traced back to Indo-European times, to include not only the Hindu god Indra, but also the Hittite sky god in Asia Minor. To all he was the rollicking, burly, irascible, red-haired and red-bearded weather or sky god. His strength was legendary (as was his appetite). He was renowned as a smiter of tremendous blows, courtesy of his hammer or doubled-headed axe, which he held in one hand; he was often represented holding a thunderbolt in the other, ready to hurl it to Earth. In his Norse guise this all-powerful god soared across the heavens in a chariot pulled by a pair of goats; a yoke of bulls transported his Hittite form, thunder rumbling in his wake. Yet despite the apparent ferocity of his demeanor, this god of thunder who thought nothing of tossing thunderstones to Earth was, somewhat paradoxically, the one most loved by people—he was a slayer of serpents and giants (using his powerful hammer) and a protector of humans. And of all his treasured possessions it was this hammer, Mjöllnir, that he most favored.[33] The word itself means "crusher."[34] With it he could smite the giants and monsters, enemies of both gods and people. In this he was helped by his magic belt, Megingjarthar, which doubled his strength, and by a pair of iron gauntlets with which he grasped Mjöllnir.

In the guise of Thor, Donar, and Thunor, this god of thunder appears to have emerged in the lower Rhineland at a time when Saxons rubbed shoulders with Celts. He was worshipped in Europe in forest clearings, his symbols being his hammer and the great oaks that covered much of central and western Europe two thousand to three thousand years ago. He was revered by the "heathen" Anglo-Saxons, who rather than wearing a cross wore his crosslike hammer as a protective amulet. In Scandinavian mythology Thor is regarded as the eldest son of the god Odin. He gradually replaced Odin in importance in people's daily lives, his most devoted followers probably being the early settlers of Iceland in the ninth century. His mother was Jörth or Mother Earth. By Viking times she had lost her importance and had been replaced by the goddess Frigg, who then sometimes became regarded as Thor's mother.

Thor's great attribute was his ability to control the weather, and he is most usually regarded as the creator of thunder and lightning. But he was more than just that. He was also the Norse fertility god. This is not as strange as it might seem: there is, after all, a close link between the weather and harvests—and not just rainfall. An old superstition has it that summer lightning ripens crops, and Thor ensured the fruitfulness of the crops and the continuity of the seasons. Even relatively recently in northern Europe old stone axes were sometimes used as fertility symbols by being placed in a hole in the plowed ground that was ready to receive the first of the spring seed. Thor then ensured the rain that makes the fields fertile and the crops bountiful. In such a way Thor, like Odin before him, can be seen as carrying on the cult of a sky god that extends back to the Bronze Age, and maybe even further back in time.

Thor's popularity stemmed from the fact that, of all the gods, it was he who fought most resolutely to defend the world against evil. It is therefore hardly surprising that he was worshipped primarily by farmers, laborers, and artisans, more so than any other god. He was the people's god. He was trustworthy and disapproved of oath-breaking, so he evolved into a god of justice and law. By ascribing the formation of fossil urchins to the hand of Thor, and because he was such a good, all-protecting god, it became obvious that the urchin was somehow lucky. It would protect people if they kept one in their pocket, or in their home. It didn't really matter whether it was placed by a door or on a window ledge—wherever the fossil was put it would protect the house from the ravages of lightning, from evil, and, in fact, from any misfortune. It was, after all, a present from Thor.

But what was it about so strange an object as a fossil sea urchin that led people thousands of years ago to link this lump of rock to the god whom they believed to rule the weather, and who played such a significant role in their lives? And is it possible to reconstruct the mythology associated with fossil urchins on the basis of folk names in conjunction with archaeological evidence? What marked the fossil as something very special is the five ambulacra that mark the surface of the smooth, round, or conical fossil. To some they might have been viewed as a five-pointed star etched, seemingly with great care, across the smooth, cloudlike surface. This is especially so of the domed *Echinocorys*. Another possibility is that the surface marks may have been seen as an allegory of Thor's hammer, Mjöllnir, and thus with Thor. Maybe during a storm, when the rain had lashed a newly plowed field and washed away some of the soil, a fossil urchin, hitherto buried from sight, was revealed. To the lucky finder of this unusual stone who else could have seared such a distinctive mark but the thunder god? And what a stone—star-crossed, with the mark of Thor's own hammer, a stone branded by Thor. He may have been the all-powerful god of thunder and lightning, but he also had another link with the heavens—with the stars that spangled the night sky. Indeed, of all the Norse gods, Thor was the one most enmeshed in stellar folklore.

In artwork from the late Middle Ages onward, Thor is often shown crowned with a halo of stars. One of the more well-known illustrations is

Olaus Magnus, engraving of the three main gods of the Vikings. On the left, Frigg; seated in the middle, Thor with his crown of stars; and to his left, Odin. From O. Magnus, *Historia de gentibus septentrionalibus* (1555), book 3.

a sixteenth-century engraving by Olaus Magnus. This shows Thor seated on his throne, and above his crown, a halo of twelve stars (and usually these are portrayed as five-pointed stars). He is flanked on his left by his father, Odin (Woden), and on his left by Frigg, the goddess who is sometimes regarded as having replaced the goddess Jörth or Earth. When these three gods are shown together, as they often were, it is invariably with Thor seated in the middle, with Odin and Frigg on either side.

The symbolism of these three gods and of the stars above Thor's head may originate from the representation of god figures as portrayed at Uppsala. It is here that a link with stars is most obvious. As sky gods, it is not surprising that as well as being associated with various aspects of the weather, some were also linked to particular stars, or to constellations of stars. Not all the constellations were attributed to Thor at the same place and at the same time. (At various times constellations were associated with different gods.) What the association does imply, though, is that even then a link may possibly have been made between the stars in the night sky and the fossil urchin with its five-pointed pattern.

Perhaps the two most obvious features in the northern sky are the constellation of Ursa Major (otherwise known as the Plough in Britain, the Big Dipper in the United States, and the Wain in Scandinavia and Germany) and the brightest star, Sirius. Ursa Major (the Big Bear) was given to various gods at different times, while Sirius, in northern European mythology, was claimed by Frigg and Freyja. Ursa Major appears to have been linked to a number of the Norse gods, Thor and Odin among them. When ascribed to

Thor, the constellation was called Karl Wagen. This name derives from the name Karl being the familiar title with which Thor was sometimes known in Scandinavia, in particular Denmark, Sweden, and Iceland. It is generally translated as "the Old Man." The term *Wain* was generally bestowed on any prominent deity who coursed the heavens in a chariot. Besides Thor, this would include Freyr, Freyja, and the old earth goddess Nerthus.

An old rhyme featuring the god figures at Uppsala describes the association of Thor with stars:

> The God Thor was the highest of them,
> He sat naked as a child,
> Seven stars in his hand and Charles's Wain.[35]

The early seventeenth-century professor of history and law at Uppsala University, Johannes Messenius, believed the seven stars in the rhyme to be the Pleiades. Others have suggested that the seven stars are instead more likely to be the constellation of Ursa Minor, and that this constellation may therefore have been thought to represent Thor's hammer, Mjöllnir. The link between Thor and stars is strengthened by the reference to "Charles's Wain." This is another name given to Ursa Major, which may therefore be Thor's chariot being drawn across the night skies by his trusty goats, Toothgnasher and Gaptooth. As Thor is often described as racing across the sky (part of the job description for the Thundergod), the stars are very appropriate symbols for him, and especially for his wagon.

Alternative explanations have been proposed for why Ursa Minor was also linked with Thor. Ursa Minor is essentially a smaller version of Ursa Major and lies much closer to the pole star than does its larger cousin. Other names for Ursa Minor are the Little Bear in England, and in Scandinavia it was once called the Throne of Thor. Maybe this is the throne that Olaus Magnus and other artists sat Thor on, as the constellation is roughly chair shaped. In the early Middle Ages chairs were unusual pieces of furniture in most dwellings and were thus a symbol of great authority. The peasants, after all, could sit on stools or benches.

Thor has also been associated with other stars. A pair of stars frequently mentioned in the Icelandic Eddic myths and called Thjazi's eyes are probably the twin stars of Gemini. In the story of Skadi, an enraged giantess demands compensation from the gods after the death of the giant Thjazi. He had been killed by the gods for stealing the apples of youth from Asgard, the home of the gods. In partial compensation for Thjazi's death Odin is said to have made two stars out of her father's eyes. However, in another account, the poem known as the Lay of Harbard, the Skadi story is explained differently. Here Thor argues that it was he who made the pair of stars:

> Strong Thjatsi, the thurs, I overthrew in battle,
> and the awful eyes of Alvaldi's son
> I cast on the cloudless sky.
> Those be the mighty marks of my great works.[36]

Another one of Thor's exploits was also said to have produced a star, known to the Norse as Aurvandil's Toe. To some this star is Alcor in Ursa Major. The story of Aurvandel was well known in Scandinavia. In Anglo-Saxon England he was called Earendel. Thor is carrying Aurvandil back from Giantland in a basket on his back. On the journey they cross a frozen, poisonous river, and one of Aurvandil's toes freezes and breaks off. Thor simply flings it up into the sky, where it turns into a star. This myth may also be linked to the Gloucestershire story of Wandil the giant. Wandil's antisocial act is to steal the Spring. But he is defeated by the gods who cast his body into the sky. He now stares down on us as the pair of Gemini stars.

As I have pointed out, Thor was not only a god of the heavens, coursing across the starry night skies with thunder crackling in his wake. He was also protector of the people who worked the land. There is a curious contradiction in that fossil sea urchins were thought to have been hurled to Earth by Thor during a thunderstorm, yet they somehow, sympathetically perhaps, acquired the symbol of luck, particularly in protecting the owner from further misguided thunderbolts sent by Thor. Seemingly of more importance was the belief that they were sent from Thor the protector, not Thor the thunderer. These fossils were sent by the god to protect people from lightning and from evil. The people knew who had sent them because the stones carried the sign of Thor's hammer. But it may also have been that at some stage this five-rayed symbol came to be also associated with the stars—with the heavens across which Thor rode in his chariot. And can we carry this echo through to the twentieth century, to the peasants of the Russian Revolution who chose as their symbol under which to fight, not only the hammer and the sickle, but also the five-pointed star— "the emblem of the reaper-peasant and the smith-worker"[37]?

But what of Mjöllnir, Thor's all-powerful hammer? In many ways, I suspect, the potency of his hammer may have been even more significant than the sign of the stars. As late as the twelfth century in Denmark it was still being carved onto headstones, in the position that was usually occupied by the rather similarly shaped Christian cross. Mjöllnir was also used at weddings to "hallow the bride." Large numbers of small iron, bronze, or even silver hammers were manufactured in the tenth century to be worn around the neck or as a bracelet or amulet. In small statues of Thor that have been uncovered from Viking age deposits, we see Thor's hammer coming out of his beard. The cross shape of the hammerhead and the bifurcating shaft that emerges from his beard are very reminiscent of the pattern of the five ambulacra incised onto the flat surface of fossil urchins like *Echinocorys*. On this surface of the urchin the five ambulacra do not make a symmetrical star-shaped pattern—the pattern is more that of a cross. In fact, the shape of the hammer amulets manufactured in tenth-century Scandinavia are so close to the shape of a cross that molds have been found in which both were being turned out from a single block—the manufacturer was apparently hedging his bets at a time of changing religions.

The potent symbolism of Thor's hammer to his followers was that it could raise up the dead to renewed life—like the Christian cross, it symbolized the power of resurrection. Hence its frequent appearance on memorial stones in Scandinavia, particularly in the ninth and tenth centuries. The power of Mjöllnir to resurrect life is recounted in one of the many stories about Thor. One night he stays at a farm and kills his trusty goats for supper. It isn't that there is a shortage of food; it is more that Thor wants to impress his host with the power of Mjöllnir. After the meal he gathers all the bones and places them on the goatskins. Raising his hammer above them, Thor instantly brings them back to life. The one hitch is that the farmer's son had, against Thor's orders, broken one of the goats' legs in order to extract the marrow. The result is that one of the goats will be forever lame. Thor is not amused.

The power of Mjöllnir to resurrect the dead is again suggested in the stories of the death of Odin's favorite son, Baldur. The great Icelandic historian Snorri Sturluson, in his thirteenth-century Prose Edda, wrote that Baldur "is preeminent and everyone praises him. He is so beautiful and bright that he glows with radiance . . . He is the wisest of the gods and the fairest spoken and the most gracious."[38] Baldur, most beloved of gods, is inadvertently slain at the hand of Odin's other son, the blind Höd. Loki, the most mischievous and evil of the gods, despises Baldur. He cuts a sprig of mistletoe and sharpens it into the shape of a javelin. He places it in Höd's hand and directs him to aim it, unwittingly, at Baldur. The shaft pierces Baldur's body, killing him instantly.

This single act is the greatest tragedy to befall the gods. At his funeral, Baldur's body is placed on a pyre on a great ship. On the body Odin lays his gold arm ring, Draupnir, which drips eight rings of similar weight every ninth night, while Thor lays Mjöllnir on the pyre, perhaps in one last desperate attempt to raise Baldur from the dead. Loki's evil act is the catalyst for Ragnarök, when the universe is consumed in a holocaust of destruction. But after Ragnarök, Baldur does indeed return from the dead, to a new, revitalized heaven. And even here Mjöllnir still has a part to play. Ragnarök, the twilight of the gods when they sink into an abyss of evil, murder, and fratricide that lead to the destruction of the universe, is replaced by a new, more beautiful world, purged from sin. And on the field of resurrection, the sons of the highest gods assemble, and in them their fathers are reborn. Magni and Modi, the sons of Thor, have brought Mjöllnir with them—not as a weapon of war, but as the instrument with which to consecrate the new heavens and the new Earth.[39]

The Christian overtones in the story of Baldur's death and the resurrection of the gods are self-evident. The transition from worshipping the old Norse gods to believing in the new Christian one was a bridge that could be crossed very easily. Thus, Mercer's broken stone axe and a fossil urchin found in the grave with burned bones may serve to illustrate a statement in the later Edda, that Thor consecrated Baldur's funeral pyre with Mjöllnir.

The reason for placing the sign of Mjöllnir, in the form of a fossil sea urchin or of a stone axe, with the remains of a dear departed soul in a grave then becomes much easier to understand. It formed the symbol of rebirth after death. It was the "pagan" cross, created in the heavens and etched onto a stone by the very god who had the power of resurrection. Another myth surrounding thunderstones that reveals their close link to Thor's hammer Mjöllnir, also recorded by Blinkenberg in Scandinavia, is the idea that after working its way deep into the ground, the thunderstone would remain there for seven years.[40] It would then find its way to the surface and reappear—it would be reborn. The parallel in Norse mythology saw Thor, who rules over summer, being deprived of his hammer in the winter. The mountain giant Thrym steals Thor's hammer and hides it eight miles (12.8 km) deep in the ground, representing the eight cold months of the year. Thor battles to get his hammer back, destroying Thrym's hall in the process. And out of the smoking heap of the ruins where once Thrym ruled,

> The spring sun rose: it shone down upon the devastated dwelling, the broken rocks, fallen stones, torn and uprooted soil, and upon the victorious god who had conquered the power of the enemy.
>
> The storm-clouds of anger were gone from Thor's brow. He stood upon the height and gazed at his work of destruction with a gentle and kindly

look upon his face. Then he called his children of men to come and instill new life into the destruction, so that farms and dwelling houses, agriculture and commerce, civic order, law and morality should arise and flourish there . . . And Thor was in the midst of them, setting up stones to mark the boundaries, consecrating the tilled land with his hammer.[41]

As Thor's hammer was brought back to life after being so long in the ground, so the finding a thunderstone may have represented the acquisition of the symbol of Thor's power. And so to take this fossil with them on their long journey into the afterlife was insurance that the dead, like Mjöllnir, would be reborn from the Earth.

Collecting fossil urchins may have been a common occurrence in northern Europe for countless millennia. But this behavior was not confined to the cold northern lands. Far to the south, in northern Africa and the lands bordering the eastern Mediterranean, other peoples had also been collecting fossil urchins for thousands of years, using them in ways surprisingly similar to those of the people in the lands of the long winter's nights.

> Never tell me that not one star of all
> That slip from heaven at night and softly fall
> Has been picked up with stones to build a wall.
>
> ROBERT FROST, "A Star in a Stone-Boat" (1923)

 # 10

Holy Urchins

Late morning, and the heat binds you to it. Your car winds down a precipitous road that plunges over 3 thousand feet (900 m) into the impressive Wadi Mujib before climbing out on the meandering approach to the Crusader's Castle of Kerak. It's the last part of a seventy-five-mile (120-km) journey south from Amman in Jordan. Pale yellow battlements tower above you in the searing midday Sun. Mad dogs, and all that. You park the car and begin the walk up the steep hill to the ruined castle. You know you should keep to the path, but the rocks on either side somehow look inviting, though you're not too sure why—perhaps they are inherently more interesting than the well-trodden path that winds toward the castle. The jagged edges of the hot, dry limestone bite hungrily at your shoes. Treading carefully between these lithic teeth, you spot a pebble, not much larger than a button—what's a pebble doing here? It is pale cream in color, and as you bend down to pick it up you see that etched delicately on its spherical surface are ten elegant rows of tiny holes, arranged in five pairs. You pick up this curious star-kissed pebble and drop it in your pocket. Suddenly you remember why you're there. The castle. These rocks are getting too hot to stay on any longer. Even the flies are keeping out of the sun. The cooler shade of the castle walls beckons. You make first for the museum inside. Down a few steps and you are enveloped by a vast, vaulted hall. It is a welcome relief—dark and cool. Countless artifacts peer wistfully from glass-topped cabinets—the personal treasures of people who lived and died here nearly one thousand years ago.

Then you see it—in a display case, nestled among other body ornaments—a necklace. Beads of many colors—green, blue, white, and yellow—adorn the fine thread. And there, at the end, dwarfing all the rest, and with a hole drilled right through its center, is what appears to be the selfsame pebble that you picked up just minutes earlier. Putting your hand in your pocket, you feel the still-warm stellar pebble. You take it out and hold it up against the one on the necklace. It's just the same, except that yours doesn't have a hole plunging right through its center. So you are obviously not the first to have collected one in these parts. Someone else had done so ten centuries earlier. You read the label in the display case and discover that the unusual pebble on the necklace is, in fact, a fossil sea urchin. Of course, limestone—once under the ocean, with sea urchins wandering across the floor of a warm sea . . .

The necklace starts you thinking. Why did someone a millennium ago pay so much attention to an old fossil? A sapphire in a necklace, now that you could understand—but a fossil sea urchin? Certainly it's interesting, but what was it that attracted someone so much that it was worn with pride in the necklace? Did it mean something very special to its owner? Did its symmetrical shape provide the lure? Or was it the star resting on the fossil's surface? And did this star hold some special meaning? Maybe a thousand years ago people liked to hang fossil stars around their neck. But did this fossil star really form the centerpiece of a necklace, or did the hole so carefully drilled through the it have some other, deeper significance . . . ?

Back on the outskirts of Amman that you left a couple of hours earlier, the wind swirls and whines around some dry, barren slopes near a busy road, where only the most recalcitrant weeds and thistles manage to eke out a parched existence. 'Ain Ghazal this place is called. The name means "Spring of the Gazelle." But it has been many thousands of years since a gazelle last drank here. This part of the eastern Mediterranean was the cradle of agriculture a little over ten thousand years ago—the place where people first learned to domesticate animals, plant crops, and put down their own roots. This event, one of the most significant for our species, marks the beginning of what we call the Neolithic period.

It was only about one thousand years after this momentous change in lifestyle that people first began to settle around the spring at 'Ain Ghazal, attracted as much by the rich animal and plant life that lived in the area as by the water. Archaeological studies begun in the 1980s have revealed the fractured, fragile bones of more than fifty species of animals that once lived in a fertile woodland dominated by oaks, tamarisks, and poplars.[1] It didn't take long for the settlers at 'Ain Ghazal to rely more and more for their food on the animals that they had learned to domesticate, rather than hunting the native animals. This must have seemed a godsend at the time—a constant supply of fresh meat, milk, and skins. But it sowed the seeds for the destruction of 'Ain Ghazal and the surrounding area. In less than a thousand years after people had settled here, the number of native species had plummeted to about fifteen, evidence of the increasing degradation of the area around the settlement.[2] Overgrazing, first by goats and then subsequently by other domesticated animals, particularly sheep, pigs, and cattle, eventually turned this wooded parkland paradise into an uninhabitable, rocky wasteland.

This destruction at 'Ain Ghazal took place in a little over 2,000 years, from about 9,250 to 7,000 years ago. At its prime it was one of the largest settlements of the region, with an estimated population of up to about 2,500 people.[3] The early phase of settlement occurred during part of what is known as the Pre-Pottery Neolithic, a time when pots were mostly made from sundried clay. The later part of the settlement includes the time of transition into the so-called Pottery Neolithic, when the concept of firing clay had been discovered. In the earlier Pre-Pottery phase, clay was more often used in a less utilitarian, but altogether more aesthetic, way. Life was good then. Plenty of food. No need to spend all day hunting for it. Time to do other things, things that had no obvious useful purpose, but in some way were good for the soul—an expression of the creative spirit.

During the archaeological excavations at 'Ain Ghazal, along with all the usual stone tools and bits of broken pots more than one hundred animal and human figurines were uncovered.[4] Most were made from clay, but others were crafted from the chalk and limestone on which the site was built. The animal figurines generally depict cattle, although there is also the occasional horse and dog. Maybe these people had an inordinate fondness for cows. Or perhaps these diminutive cattle were made for a purpose, either ceremonial or ritual.

The clay cattle were, for the most part, simple, and all looked alike, emphasizing the head of the animal. The human figurines, however, were more diverse. Some depict the entire body, while others are just dismembered heads or torsos. Most distinctive are a few that have distended abdomens and pendulous breasts that have been interpreted, not unreasonably, as fertility figurines. But perhaps the most stunning piece is one not made from clay at all, but skillfully carved from limestone. Standing five and half inches (14 cm) high, this elegant statuette of a female accentuates most emphatically her pregnant condition.[5] Chest and lower limbs are minimized, but enormous arms taper to tiny limbs when they reach the stomach.

Plaster was also sometimes used for making human statues. Reaching about three feet (1 m) high, these startling creations stare out from eyes with irises made from bitumen, deposits of which occur in and around the Dead Sea.[6] Bitumen was also used as an eyeliner to further accentuate the piercing stare. To make the bodies even more lifelike, some had a pink "wash" applied to them. The use of such coloring was not confined to the human figures. The floors of some rooms in the settlement were sometimes covered with plaster and burnished with a red pigment. This practice was carried out for much of the two-thousand-year period of occupation of the site.

These early Neolithic works of art, expressions of the innate creativity of these people, form only a very small part of the manufactured items excavated at 'Ain Ghazal. Thousands of stone and bone tools have also been found, along with countless fragments of bowls and jars. As the archaeologists painstakingly recorded every shard of pottery and splinter of stone tool during their excavations, other strange objects occasionally turned up—objects to intrigue the archaeologists who found them and to provide a fleeting break from the monotony of finding yet another fragment of clay pot; objects gathered, perhaps, to assuage the hunting urge. Among these were four fossil sea urchins. But there was something altogether unusual about these particular fossils. All showed very obvious signs of having been modified and used by the inhabitants of 'Ain Ghazal.[7]

Finding fossil sea urchins in a site located on fossiliferous Late Cretaceous limestones is really not too surprising. This rock around Amman forms part of the 70 million–80 million-year-old Late Cretaceous chalky limestones of the Ajlun Group and covers about two-thirds of present-day Jordan. Like the chalks of northern Europe that were deposited at much the same time, the limestones contain many fossils. But whereas the chalk hills in England over which Maud roamed contain the heart-shaped *Micraster* and the helmet-shaped *Echinocorys*, different fossil urchins are found in Jordan. The two most common types in the Jordanian chalks are a heart urchin called *Mecaster* and a small, circular one known as *Coenholectypus*. The *Mecaster* not only sounds like *Micraster* but also looks like it too. *Coenholectypus*, on the other hand, is a more primitive form of irregular urchin, belonging in a group known as the holectypoids. While this type of urchin is occasionally found in the English and northern European chalks, it occurs far more frequently in the Jordanian chalks. When viewed from above, regular urchins have their anus on top of the calcareous body, and their mouth centrally positioned below (imagine standing on your head and kissing the ground—you'll be doing a pretty good impersonation of a regular urchin). *Coenholectypus*, though, is classified as an irregular urchin because its anus is tucked underneath the lower surface, not too far from its mouth. This is still centrally positioned (and I'll leave you to imagine the contortions needed to conjure up the human equivalent of this one).

It wouldn't be unreasonable to suggest that the odd fossil found in limestone country, even if it did come from an archaeological excavation, found its way there by natural means, simply by the natural process of weathering out of the limestone. Who can say how many fossils found in archaeological sites all over the world have been discarded because their finders made just such an assumption? Worthington Smith might have thought, *Mmmm, what a coincidence that Maud and her baby were buried in a very dense concentration of fossil urchins*, then left the fossils behind with not another thought. Fortunately, like a few other meticulous archaeologists such as Pitt-Rivers and Curwen, Smith carefully collected and recorded everything.

But with the four urchins found at 'Ain Ghazal, it was all too easy to see that someone had collected them long before the archaeologists found them. All show very obvious signs of human modification.[8] The most striking of the four urchins is little bigger than a button and much like a hamburger bun in shape. The five ambulacra stand out as deep red scars on a salmon pink surface. But there are many strange features to this particular fossil. On close examination of its domed upper surface it looks as though somebody has picked at it with a sharp instrument, revealing flecks of white calcite beneath the pink, stained surface. Not only that, but this picker spent more time on the five ambulacra themselves. Much of the white crystalline structure has been carefully scraped out, revealing the deep red surface beneath. The pink wash on the plaster figures; the red burnished floors of their dwellings; and now someone had very carefully brought out the red coloration of part of these fossils. More than that, this person had been very keen to highlight the star-shaped pattern of the ambulacra. But these mere decorative tinkerings are nothing when compared with the main modification of this and the other three fossils.

Coursing unwaveringly through the fossil from top to bottom is a hole. And a very meticulously drilled hole at that. Starting on the lower surface right in the center, where in life the urchin's mouth would have been located, the modifier of the urchin started the long, careful process of drilling a hole. As with most urchins that have had holes drilled through them (and they have been found in deposits from the Paleolithic through to the post–Iron Age in many of the countries that rim the Mediterranean), the hole is tapered, narrowing toward the center of the fossil. From the shape of the hole it is possible to establish the nature of the instrument that was used to make it.

In most cases it was probably a narrow, sharpened tool, like an arrowhead. In this 'Ain Ghazal urchin the hole on the lower surface takes up nearly one-third of the width of the entire urchin. But as it passes through the urchin it narrows to about half this width. Yet there is something very odd about this hole. Although it emerges on the upper, domed surface, it doesn't come out in the middle. No attempt had been made to drill from the upper surface to produce the usual hourglass-shaped hole—this one has just a small, neat hole set a little off center. Usually, with perforated urchins from Neolithic and Iron Age sites in Jordan, as well as the many found in other countries around the Mediterranean, the hole goes straight through, from center to center.[9] It tapers from both directions, as it was clearly made by drilling from both sides, to ensure this central position. Like the urchin on the necklace at Kerak Castle, the hole through the 'Ain Ghazal specimen may have been made so that the urchin could be worn as an ornament on a necklace or bracelet. Some urchins altered in this fashion may have been used as spindle whorls. However, as I discuss below, the hole may have carried altogether a much deeper significance.

So what went wrong with the urchin fossil from 'Ain Ghazal? One possibility is that the person who drilled it wasn't very accomplished and just gave up after realizing that the hole was off center. On the other hand, given the position of the hole on the upper surface, nestling perfectly between a pair of ambulacra, it may be that the crafter was very good indeed. Let me explain.

Forget for the moment the things that we call stars which pockmark the heavens, and consider only the five-rayed pattern that we also call a star and which straddles all urchins (and which in all truth looks nothing like the point of light in the night skies that we also call a star). We know from their clay, limestone, and plaster figurines that the talented artists of 'Ain Ghazal were fascinated by the human form and its creative representation. Just ponder for a moment what they would have made of the five-rayed pattern on these fossils. In this delicate (and to them probably mysterious) feature, might they not have seen their own likeness—head held high to the sky, arms outstretched to embrace the world, and legs spread apart?

The sea urchin *Coenholectypus larteti*, carefully bored through the posterior pair of ambulacra and possibly used as a fertility object; from the Neolithic (Middle PPNB) site at 'Ain Ghazal, Jordan. Fossil diameter 0.6 inch (1.6 cm). Author photo.

But not standing firmly on the Earth. As Leonardo da Vinci showed in his drawing of the so-called Vitruvian Man—a classic demonstration of the relationship of the outstretched arms and legs of the human form to geometric shapes—with legs together and firmly planted on the ground, and arms held out sideways, the tips of the appendages and the head touch the sides of a square. However, with the legs spread and the arms held apart, but pointing upward, the two pairs of appendages, with the head, form the points of a star, around which can be inscribed a circle—the same pattern that adorns the circular fossil. Could these early Neolithic people have recognized this relationship many millennia before Leonardo, the idea first presenting itself in the form of a little fossil sea urchin? For people striving to represent the human form in clay or stone or plaster, here, maybe, was a ready-made representation, and one which could become, if suitably modified, an object of even more potent symbolism.[10]

But what of the hole? Set slightly off center of the star, it takes on a whole new meaning if the star is thought to represent the human form. Like their own hand-crafted human and animal figures, the perforated urchin can, without too much difficulty, be interpreted as another fertility symbol. Irregular urchins have four ambulacra that form a mirror image of the other, and a fifth, single one. If the single ray represents the head, then the pair of ambulacra below it are the arms and the pair at the base, the legs—arrayed in Leonardo's "circle" pose. The perfectly circular hole drilled into the red and white urchin lies perfectly at the junction of the two legs—the most potent of all fertility symbols.

Although appearing some thousands of years later, a similar representation can be seen in carvings of the Egyptian goddess Nut. One, on the schist sarcophagus of Princess Ankhnesneferibe (Twenty-Sixth Dynasty, circa 525 BC, Thebes), shows the goddess with arms held high above her head, her legs slightly apart, and a large circle over her genitalia. In Heliopolitan theology Nut was the daughter of the air god Shu and sister of the Earth god Geb. She personified the vault of Heaven. The heavenly bodies were her children and each day they were said to enter her mouth and emerge again from her womb. She is also known as the mother of the Sun god, Ra, whom she swallowed every evening and to whom she gave birth each morning. Five-pointed stars often spangle her image—but more on that in the next chapter.

As depicted in stone sculptures, intercourse and pregnancy played a significant role in ancient Near Eastern mythology, to the point that it permeated it. Childbearing was often used as a metaphor to depict the mysteries of the origins of the universe, the planets and stars. The concept was that everything, whether animal, plant, mineral, or star, was the product of sexual conception. Thus, Nammu, the sole primeval Mesopotamian deity, was thought to have given birth to An-Ki, who were sky and Earth. An-Ki, in turn, were seen as the inseparable pair who then begat water, wind, the Sun, the Moon, and all the stars. Nature's bounty was also perceived as being renewed each season by divine sexual intercourse. Ki, for example, was thought to be responsible for all the vegetation on Earth by sleeping with An. In Sumerian/Babylonian times the deity Inanna/Ishtar caused the seasonal germination of plants and the propagation of flocks in spring by having intercourse with the god Dumuzi. Denise Schmandt-Besserat, who has interpreted the 'Ain Ghazal stone statuette of a pregnant female as a fertility symbol, has suggested that this particular early Neolithic work of art embodies a mythical female that engendered fertile crops.[11]

The Mesopotamian myths, while recorded on cuneiform tablets about 4,500 years ago, have a deeper history that extends far back into the prehistoric oral tradition, perhaps as far back as the early Neolithic. Such myths concerning the origins of life and the mysteries of its rebirth each spring were intricately linked in many societies with the day-to-day experience of sexual intercourse and the formation of life. If the settlers at 'Ain Ghazal

The sea urchin *Rachiosoma larteti*, bored through its center and probably used as a spindle whorl; from the Neolithic (Middle PPNB) site at 'Ain Ghazal, Jordan. Fossil diameter 0.8 inch (2 cm). Author photo.

saw the human form in the fivefold pattern on the urchin, then using the fossil as a means of helping to ensure continued fertility of the people, their flocks of sheep and goats, and the soil for a bountiful harvest would have appeared quite natural to them.

In addition to being discovered in Neolithic deposits in the eastern Mediterranean region, perforated fossil urchins have been found further south, in Sudan,[12] Niger, Libya, and Algeria,[13] indicating an interest in fossil urchins no less intense than in northern Europe. However, a number of younger Iron Age sites in the Near East, particularly Jordan, have yielded relatively large numbers of these fossils, with many showing evidence of various types of artificial alteration. It looks as though they were more useful to these people in life than in death, for none have been found associated with burials. Much of this Iron Age material has come from Busayra in southern Jordan, and was collected by the archaeologist Crystal-M. Bennett between 1971 and 1974. Busayra is one of only three sites investigated thus far in an area occupied by the Edomites between the eighth and second centuries BC.

In Old Testament times the Edomites were the neighbors of the Israelites,

Moabites, and Ammonites. The biblical kingdom of Edom extended south from the southern end of the Dead Sea almost to the Red Sea, and stretched east to the Syrian desert. One of the reasons for Crystal Bennett's excavations at Busayra was to try and establish whether this was the site of the biblical city of Bozrah, the capital of Edom for some time. Passages in the Bible (Numbers 20:14–21) tell of the king of Edom refusing passage through his land to the Israelites when they came out of Egypt. This is said to have occurred at around 1230 BC.

Bennett's excavations and subsequent work led by Piotr Bienkowski indicate that the Busayra site is not the likely location of Bozrah, for there is no substantial evidence that the settlement existed in the thirteenth century BC. The earliest dates, on the basis of the pottery recovered, are late eighth century. The youngest are about 200–300 BC.[14] During Crystal Bennett's excavations a total of seventy fossil sea urchins were recovered. Half of these came from the "Acropolis," an area consisting of palatial and temple buildings. The rest of the fossils were found in areas that were thought to have been occupied by ordinary domestic buildings.[15]

Only twenty-four urchins come from stratified contexts, but these show that specimens were being collected throughout the roughly five-hundred-year occupation period of the site. Significantly, a number of urchins were found inside what have been interpreted as "temple" buildings, suggesting that something more than mere idle curiosity was attached to these fossils by the Iron Age inhabitants of Busayra. Both regular and irregular urchins were collected by the Edomites. Only one species of the regular variety can be identified with any degree of certainty, an urchin called *Heterodiadema lybica*, common in Late Cretaceous limestones in Jordan. Most of the urchins found at Busayra, though, are irregulars, with a third being the heart urchin *Mecaster* and the remainder the holectypoid *Coenholectypus*, the same as the fertility urchin from 'Ain Ghazal. And nearly five thousand years after an inhabitant of 'Ain Ghazal picked carefully at his urchin to enhance the star pattern, Edomites were doing much the same thing, to the same type of urchin.

On one urchin, the surface of the test has been stripped off using a sharp tool, revealing a fractured, cleaved surface of calcite. Because the cleavages radiate out from the center of the urchin, they enhance the creature's radial pattern[16] ordinarily expressed by the five ambulacra. In this specimen there has been differential mineralization during the fossilization process, with red, iron-rich calcareous ambulacra between white, pure calcite interambulacra. Removal of the surface of the urchin test has enhanced this color differentiation, producing exactly the same effect as in the 'Ain Ghazal urchin. This practice seems to have had a particularly long heritage. Another modified urchin from Busayra also experienced a similar partial stripping of its upper surface. But it has also been abraded by persistent rubbing on a coarse surface, producing a slightly flattened lower surface.

Such scraping and rubbing of the fossil to produce an enhanced, more pleasing visual effect are seen in yet another urchin, one found in the Amman Citadel. The fossil, a regular urchin called *Raphiostoma rectilineatum*, was found in Islamic (Fatimid) levels (909–1171 AD) in an excavation carried out by Crystal Bennett in 1979. The effect of the abrasive rubbing of the urchin was to flatten off the otherwise convex upper and lower surfaces.[17] It is not clear whether this alteration resulted in the fossil serving a useful function. One possibility is that the modification enabled the fossil to be used as a gaming piece.

The other human modification of these fossil urchins that has a similarly long heritage is the practice of boring a hole through them. In addition to the four such fossils known from the Neolithic of 'Ain Ghazal, similar examples have turned up in a number of Iron Age sites in Jordan. Two specimens of *Coenholectypus* found during archaeological excavations at Busayra were artificially perforated. Unlike the 'Ain Ghazal fertility urchin, however, these holes had been drilled right through the center of the fossils in the same way as the other three 'Ain Ghazal urchins. *Coenholectypus* lends itself well to such treatment, because the mouth is situated on the lower side and is sunken. Similar perforated urchins have turned up in a number of other archaeological sites in Jordan, indicating that this was a relatively widespread practice. Early Iron Age (ninth and eighth centuries BC) excavations at Tell Jawa, for example, yielded three urchins, two of which were perforated.[18] The beveled holes were drilled from both sides. Younger, late Iron Age (late eighth and seventh centuries BC) deposits also produced an urchin perforated in a similar fashion, plus a broken heart urchin.

Fossil urchins in archaeological sites of the eastern Mediterranean are not restricted to the mainland, though. Some have turned up on the Mediterranean islands of Cyprus, Crete, and Malta. A number of these finds support the idea that fossil urchins possessed some sort of spiritual significance. Certainly, their frequent occurrence in deposits representing the remains of buildings considered to be temples or other places of worship tends to support the view that these were votive objects. For example, in an early Neolithic site at the Cape of Apostolos Andreas-Kastros in the area of Rizokarpaso, northeastern Cyprus, a Pliocene urchin, *Sphaerechinus granulosus*, was found.[19] And at Kommos in southern Crete a fragment of fossil urchin was found in Bronze Age Late Minoan II deposits.[20] From two

Iron Age levels on the same island (circa 760–600 BC and 375 BC–AD 170, respectively) two urchins have been collected.[21] Indeed, many fossils are known from Iron Age deposits at Kommos (bivalves and gastropods as well as the urchins), and it is thought that often they had been placed in sanctuaries as offerings to a deity.[22] But it was a find on the island of Malta in the early twentieth century that showed conclusively that fossil urchins were spiritually of great importance.

It was early morning on the twentieth of July, 1915, and Themistocles Zammit was standing knee deep in verdant crops in a field in Tarxien, Malta. He was probably asking himself what he was doing there. Chasing wild geese was one thing that perhaps sprung to his mind. What may have been frustrating him was that the field was so close to an archaeological site that he had been working on for some years. It was called the Hypogeum, and there was still so much left to do before the excavation was completed. The Hypogeum was an underground temple built about five thousand years ago. Zammit had been excavating it since its discovery in 1902, and had so far recovered more than seven thousand skeletons.[23] Some had been buried with shell necklaces, polished stone amulets, and funeral pottery.

Still, he had been putting this day off for two years, and just couldn't put it off any longer. Once he got it sorted out, he could get back to work. In 1913 the farmer who owned this particular field had turned up at the museum where Zammit worked and which he had founded some years earlier. Excitedly he informed Zammit that he had been digging in one of his fields a little deeper than usual when he had struck some large blocks of stone. Not wishing to sound too uninterested, Zammit suggested that the farmer return to that area, dig a trench, and then see what he could find. Not expecting to hear from him again, Zammit was taken a little aback when the farmer returned a couple of weeks later to report that after removing just under three feet (1 m) of the whitish soil, a square block of stone had appeared, followed soon after by another. They certainly were not natural. And along with the stones came piece after piece of broken pottery. Yet another site that Zammit would have to excavate.

However, it only took a week of digging for Zammit to realize that beneath this field at Tarxien was a significant megalithic building. Careful removal of the surface soil initially revealed layers of Roman age. As his workers methodically dug deeper a larger area was reached of black ash and broken pottery. These pottery fragments subsequently turned out to be the remains of funerary urns, along with some burned human remains dating from the Bronze Age. Excavation finished in September and recommenced the following April. By the end of the September of the second year of digging, most of the walled spaces had been cleared of soil and debris, revealing a series of elliptical rooms. Zammit recognized two distinct buildings. By the end of the following field season a third had been found. All had been used as temples, Zammit concluded. The three covered an area of about 8,750 square yards (7,000 m^2). Columns and pillars had been revealed, and many were engraved with intricate spiral decorations as well as carvings of bulls. Massive stone basins up to almost a yard (1 m) across were also unearthed.

There was one particular pillar that Zammit called the Jeweled Pillar, within which five shallow holes, situated about one foot (30 cm) above the floor, had been excavated for the worshippers at these Tarxien temples. Amazingly, offerings placed reverently in these little enclaves thousands of years before were still there—some dark, polished stones and two fossil sea urchins. Zammit was in no doubt that these were venerated by the worshippers at Tarxien, and he believed that, regarding the urchins, "the delicate character of the ornamentation speaks volumes for the meticulous care and artistic sense of these primitive people."[24]

To Zammit, the presence of many Neolithic temples on Malta, such as those at Tarxien, indicated a thriving center of religious activity. Five thousand years ago even fossil urchins played a part in the daily and seasonal rituals of the people of these islands. To Zammit,

> The numerous megalithic sanctuaries of the Maltese group indicate a sturdy primitive stock of people, peaceful and religious, who turned their hands to the erection of buildings to honour the unknown powers that held them in subjection. In these temples, the navigator who approached the sheltered harbours offered sacrifices to propitiate the powers of the sea during dangerous crossing. Malta was the holy island of Neolithic faith, the half-way house of the early mariners, who trusted themselves to their frail wooden craft, full of hope in a protecting power.
>
> The mariners, guided by the natives, hied to the nearest sanctuary, where a fat bull or a humble goat was bled before the altar and burnt before the holy image. The oracle predicted a prosperous journey or the fulfillment of hope and dreams.[25]

Support for the use of fossil urchins as either spiritual or magical objects comes from other places in Jordan and Israel. One is a small Iron

Age shrine recently discovered by Michèle Daviau and her team at an archaeological site in Jordan called Wadi ath-Thamad. Of the four urchins found here, three show signs of artificial alteration. Attempts were made to drill holes in two of them, but they were never completed for some reason.[26] The other site where fossil urchins have been found that may have been used in spiritual or ritual activities is the late Bronze Age temple at Timna, near Eilat in southern Israel. Three specimens of the regular urchin *Heterodiadema* and two *Coenholectypus* were found here.[27] The Canaanite site of Timna, situated in the Aravah Valley, exhibits Egyptian influence in its layout and construction. Copper mining activities were carried out by Egyptians here during the Nineteenth and Twentieth Dynasties. The temple and associated mining activities continued to flourish until the period of Ramses V. A well-marked route existed from Egypt through the Sinai Desert and southern Negev to Timna. This is illustrated by cartouches of Ramses III on rocks at these intervening sites. The temple at Timna was dedicated to the Egyptian goddess Hathor. Kenneth Oakley suggested that fossil urchins would have proved to be suitable offerings, for Hathor was goddess not only of the underworld, but also of the night sky, the "star-like ambulacral areas suggest(ing) the night-sky."[28]

There is evidence that early Egyptians were also in the habit of collecting fossil urchins. The Egyptian Tomb 361 at Tell el-Ajjul in Gaza, excavated in the 1920s by Sir Flinders Petrie, produced a perforated urchin, again probably *Coenholectypus*.[29] Nine pierced fossil urchins of the Tertiary genus *Plagiochasma* were found in a Chalcolithic (late Neolithic) tomb at Toukh in the late nineteenth century.[30] These silicified urchins were thought to have formed part of a single necklace. The only evidence of fossil urchins having been placed in more simple graves comes from an early Predynastic grave, about 5,300 years old, from Naqada. Although the body was missing from one particular grave, it did yield a fossil heart urchin called *Linthia* cf. *cavernosa*.[31] Given that people have been undoubtedly collecting fossil urchins in the Middle East for at least ten thousand years, it may well have been that the perceived powers attributed to these objects were passed on to the characteristic five-pointed star that they carried.

There is another fossil urchin known from an archaeological site in Egypt, one that may hold the key to understanding why this five-pointed symbol came to be associated with stars. This single specimen carries its own unique story—for we know the name of the person who found it over three thousand years ago.

My limbs are the Imperishable Stars. I am a star which illuminates the sky.
Utterance 570 from the Pyramid Texts

11
The Morning Star

On the eastern side of the river Nile in Egypt, largely consumed by the slowly creeping sprawl of Greater Cairo, lie the remains of the ancient sacred city of On. Known to the ancient Greeks as Heliopolis, little remains today of the legendary splendor of this city, save for a lone obelisk standing like a solitary finger, pointing forlornly to the stars. The obelisk, which commemorates Senwosret I, a powerful pharaoh of the Twelfth Dynasty (1965–40 BC), is the only one that remains after centuries of pillaging the many that once stood here. Two raised by Tuthmoses III of the Eighteenth Dynasty (1479–25 BC) were removed by the Romans in 12 BC and taken to Alexandria, where they were placed in front of a temple dedicated to Augustus Caesar. But this was not to be their last resting place. In 1878 one known today as Cleopatra's Needle was shipped to England and re-erected on the Victoria Embankment in London. The other found its way to Central Park in New York.

Few who pass by these towering obelisks today can imagine their spiritual significance to the powerful priests of Heliopolis 3,500 years ago. These priests were custodians of the secrets of initiation into the state cult that centered on the temple of Ra, the Sun god. There is much evidence to indicate that these priests were also, for all intents and purposes, astronomers, as their religious cult centered not just on the Sun, but also on the stars.

In the early Dynastic Period, up to five thousand years ago, Heliopolis was already a religious center, although it appears to have consisted of little more than a crude sacred pillar for which it was named. Its Egyptian name, *jwnw*, meaning "pillar," was pronounced "a-na," but transcribed as "On." By the time the great pyramids at Giza were built, about 4,450 years ago, Heliopolis had become the religious heart of a united Egypt centered on astronomical worship, with the high priest being known as the Chief of Observers. A temple had been erected here dedicated to the god Re-Atum, the Complete One, the father of all the gods. Whereas in later dynasties he became identified with the Sun god Ra, during the first few dynasties, when the pyramids of Giza were being built, Re-Atum was the One God, and the creator of the Sun and all else within the world.

In Egyptian mythology, creation began with the Earth as a mound that had emerged from Nun, the personification of the primordial ocean. On this mound grew a lotus, from which Re-Atum appeared, bringing life into being. Re-Atum's offspring were Shu, the air god, and Tefnut, the moisture goddess. The issue of this pair were Geb, the Earth god, and Nut, the sky goddess. These two mated, even though they were interrupted by Shu, their father, who tried all he could to keep them apart. Shu, as air, came between them, lifting the canopy of the sky away from the Earth, thus parting the lovers.[1]

Despite this, Nut managed to produce four anthropomorphic gods who dwelled on Earth: the gods Osiris and Seth, and the goddesses Isis and Nephthys. These nine formed the Great Ennead of Heliopolis. Osiris and Isis form the basis for one of the greatest of all myths of ancient Egypt: the divine couple who first ruled Egypt. They produced a single son, Horus. Osiris was frequently identified with the phoenix, which, along with a sacred stone, may have symbolized his generative powers that created Horus from the womb of Isis. In earlier times these three, Osiris, Isis, and Horus, were represented by Sah, Sophet (Sopdet), and Sopdu (Soped), respectively.[2] And it is the name Sopdu that has turned up, inscribed as a hieroglyph many thousands of years ago, on the most unlikely of all objects—a fossil sea urchin. We don't know exactly where he found the fossil, but to its finder, a priest at Heliopolis, this urchin was something very special. And about 3,500 years later, an Italian archaeologist found it again.

In 1903 Ernesto Schiaparelli (1856–1928) began a series of expeditions that was to take him during twelve seasons of digging to some of the most significant archaeological sites in Egypt. Appointed director of the Museo Egizio in Turin in 1894, Schiaparelli established a major collecting policy for the museum, focusing on ancient Egyptian artifacts. As head of the Italian Archaeological Mission in Egypt (which had been created with funding provided by the king of Italy, Victor Emmanuel III), Schiaparelli worked at classic sites such as Giza, Hermopolis Magna, El Hammamia, Asyut, Aswan, Kaw el Kebir, and Heliopolis, making important discoveries at all these places. He excavated in the Valley of the Queens, and in the process discovered the plundered tombs of the sons of Ramses III, Khaemwaset and Amenhirkhopshef, and the tomb of Queen Nefertari, one of the principal wives of Ramses II. His most significant discovery, though, was probably the intact tomb of the architect Kha and his wife in the necropolis at Deir el Medina.[3]

The fabulous items that Schiaparelli discovered, collected, and transported back to Italy make the Egyptian collections of the Museo Egizio one of the most significant in the world. But among all these wonderful objects there is one that, given the wealth of material that he and his fellow workers uncovered, is seemingly incredibly mundane—a fossil sea urchin. During the excavations it is likely that many such fossils would have been observed, but generally not collected. After all, fossils such as these occur relatively frequently in the 40-million-year-old Eocene limestone outcroppings in many parts of Egypt that were used in building many of the great monuments, including the pyramids at Giza.

Maybe it was just a trick of the light—a specimen pushed over with the foot, perhaps, to reveal its flat, lower surface; perhaps it was also early in the day, or maybe a little before sunset, when the low rays of sunlight raked across this fossil urchin, picking out the most delicate surface details. Whatever the exact scenario, Schiaparelli would have seen the pentagonal mouth of the urchin in the center of a flat, oval surface; radiating out from this were the five ambulacra, making the pattern of a five-pointed star. But what clearly would have attracted Schiaparelli to this particular starcrossed stone was the fact that somebody, long ago, had written on it. Running in a circular pattern are twelve hieroglyphs, each about four-tenths of an inch (1 cm) high, carefully engraved and well executed.

The fossil was one of the objects uncovered by Schiaparelli during excavations at Heliopolis between 1903 and 1906. The remains of an openbrick building of circular plan were uncovered at that time, but due to opposition from the landowners the team was unable to continue their

Ventral surface of the fossil urchin *Echinolampas africanus* and found, according to the hieroglyphs inscribed on its surface, "in the south of the quarry of Sopdu by the god's-father Tja-nefer" in about 2000 BC. Fossil length 2.4 inches (6 cm). Photo courtesy of Museo Egizio, Turin, Italy. Used by permission of Ministero per i Beni e le Attività Culturali-Soprintendenza per i Beni Archelogici del Piemonte e del Museo Antichità Egizie.

excavations to define the building completely. Along with the fossil, Schiaparelli found a wide variety of other objects, such as statues and vases, some of which dated back to the Old Kingdom (2160–2690 BC). Perhaps this eclectic collection once formed part of an ancient museum.

The inscribed fossil was identified as a cassiduloid urchin called *Echinolampas africanus*, a relatively large species characterized by a high domed surface, across which run five prominent ambulacra, as well as the five-rayed star that radiates out from its mouth on the lower surface. And it was on this lower surface that the hieroglyphs had been inscribed. It wasn't until 1947, however, that the hieroglyphs were first translated, by Ernest

Scamuzzi.[4] Schiaparelli had not translated them, mainly because of his difficulties with interpreting the fourth and fifth hieroglyphs. Scamuzzi, though, was able to make them out. What was written on the fossil would gladden the heart of any museum curator, as it held to the curatorial maxim: always record where the specimen was found and by whom. And that's just what the inscriber of the hieroglyphics did, making this in all probability the world's oldest curated specimen, perhaps from the world's oldest museum collection. According to Scamuzzi's translation, the fossil tells future generations that it was "found south of the stone quarry of Sopdu by the 'Father-of-God' T-nofre."

Scamuzzi believed that the names Sopdu and T-nofre (more recently transliterated as "Tja-nefer"[5]) both provide an indication of when the urchin was inscribed, and therefore when it was likely to have been found. He argued that the spelling of the god Sopdu's name is similar in style to that found in the Sinai mining area, particularly like examples from the reign of Amenhemet III and Amenhemet IV (Twelfth Dynasty, about 2060–1990 BC). The priest who found the fossil, Tja-nefer, possessed a name that was widespread in ancient Egypt from the New Kingdom onward. His priestly office can be described either as "Father-of-God," "Father divine" or "God's-father."

More recently, in her book on the god Sopdu, Inke Schumacher reintepreted the hieroglyphs found on the fossil. Transliterated they read: gmi Hr rsy ikw %pdw in it-nTr *A-nfr,

The hieroglyph on the fossil urchin found by Ernesto Schiaparelli at Heliopolis, Egypt. This translates as "Found in the south of the quarry of Sopdu by the god's-father Tja-nefer."

This is best translated as "Found in the south of the quarry of Sopdu by the god's-father Tja-nefer."[6] So, rather than having found the urchin "south of the quarry," it was more likely to have been found *in* the southern part of the quarry. The question is, just where was this quarry?

Scamuzzi believed that clues to its location come from the mention of Sopdu, the god honored together with King Senfore of the Fourth Dynasty and the goddess Hathor in inscriptions in the Sinai mining area, and spelled in the same way as on the fossil. In the Sinai region Sopdu was called "He who strikes against the Asians" and "Lord of the East," as well as "Master of the Oriental foreign people." However, in terms of the geology, this provenance of the fossil urchin is unlikely, because few limestone quarries occur in the Sinai. Schumacher has suggested that the quarry lay on the eastern side of the Nile, at either Gebel el-Ahwar or Tura. She favors the latter, because a stele inscribed with the date of Year 2 of King Amenhotep of the New Kingdom was found in Tura quarry when a new section was opened up. The stele depicts the god Sopdu, along with Anubis and Werethekau. Whether or not this was the place where it was actually found, the fossil was obviously so precious to Tja-nefer that he not only recorded on it the details of where it was found, but took it all the way back to Heliopolis with him.

Coincidentally, around the time that Maud and her child were being laid to rest in a wreath of fossil urchins on a windswept downland in southern England, an Egyptian priest from the holy center of Heliopolis was collecting an urchin, and taking special care of it. But what was so special about this fossil that Tja-nefer was compelled to keep it and take it back to Heliopolis? Could it have held as much spiritual significance to him as the fossil urchins did to Maud and her people? And what was Tja-nefer doing at this quarry in the first place? Had he gone there because he knew it to be a place where such strange stones could be found? Or was it merely a chance encounter? In Tja-nefer's mind, was his finding ordained by the stars? Then again, perhaps he was just another in the long line of eccentric fossil urchin collectors. I can quite understand his decision to keep the fossil if he was. But for him to have then gone to the trouble of inscribing, or having someone else inscribe for him, both his name and the place where the urchin was collected points to its having meant something quite special to him.

I suspect that what might well have appealed to Tja-nefer most was that here in the hot, dry Sinai desert he had found, engraved on a rock, perfectly symmetrical five-pointed stars. With this particular urchin, *Echinolampas*, the star pattern is clear not only on the domed upper surface, but on the lower surface as well. Of all pre-Christian peoples, it was the early Egyptians who appear to have been most taken by the symbol of the five-pointed star, or *saba*, as they called it. To them the pattern was synonymous with the stars in the sky. This was known as the Duat, and was represented hieroglyphically by a five-pointed star set within a circle[7]—very reminiscent of the fossil sea urchins they would frequently have found in their limestone quarries.

From the time of the Old Kingdom onward, the Egyptians believed that mortals could be reborn as stars that would be continually visible above the horizon. To their eyes the sky was a celestial ocean that encircled the world. The womb of the sky goddess Nut was the underworld into which the Sun went each night through her mouth, to be reborn the next day.[8] And being the night sky, Nut's body was swathed with stars. On the 3,200-year-old sarcophagus of the pharaoh Merneptah, she can be found clothed only in a constellation of five-pointed stars, as though spreading herself over the deceased so that he might be placed among the imperishable stars and thus gain eternal life. At the center of each star lies a small circle. In the same way, Nut arches across the entire ceiling of the elaborate tombs of the pharaohs Seti I (1294–1279 BC) and Ramses VI (1143–1136 BC), in the Valley of the Kings. And flanking her outstretched body is a celestial aura of five-pointed stars. In some depictions the stars fill her body[9]—and from her body came forth Osiris, Seth, Isis, and Nephthys, and from the union of Osiris and Isis issued the fifth, Horus—the Morning Star. Such symbolic stars must have meant so much at Heliopolis that even the chief priest wore a cloak covered in them.[10]

So why did the Egyptians choose to represent a star as having five points? Why not six or seven? The answer may lie in the five gods, Osiris, Isis, Seth, Nephthys, and Horus, the first four of whom belonged to the Great Ennead. The Egyptians divided their year into twelve months, each consisting of three ten-day weeks.[11] This left five spare days at the end of the years. These were called the epagomenal days, and each was designated as a birthday of the gods Osiris, Isis, Seth, Nephthys, and Horus. Given the Egyptian priests' astronomical obsession, it is possible that each ray of a five-pointed star represented one of these gods. Certainly, the use of the five-pointed star symbol in hieroglyphics was usually associated with aspects of time, such as in the words *morning* and *month*. Five-pointed stars also appear on an "hour table" that formed part of a star clock inscribed on the tomb of Ramses VII.[12] And this same five-pointed pattern found on the commonly dome-shaped stones peppering the Egyptian limestones would only have reinforced the deep symbolism of this pattern and confirmed the fossil as a token of the gods.

The early Egyptians believed that at the edge of the star-spangled night sky was the Field of Reeds, sometimes known as the Field of Offerings (the Elysian Field to the Greeks), where the stars never set and to where the souls of the dead traveled.[13] These stars played a particularly significant role in the pyramid alignment ceremony undertaken by astronomical priests during the second year of a pharaoh's reign, following the burial of his predecessor. This ceremony, known as *pedj shes* (stretching the cord), involved the choosing of a site for the new pharaoh's pyramid. Of critical importance in the ceremony was the alignment of the pyramid to the cardinal points. By using the simultaneous transit of two of the circumpolar stars in the Orion and Great Bear constellations, an accurate direction of north could be determined. In this way, the alignment of the pyramid was achieved with an astonishing degree of accuracy.[14] And it was to Heliopolis, the headquarters of these astronomical priests, that Tja-nefer took his star-studded fossil urchin.

Much of our understanding of the beliefs, rituals, and spirituality of the early Egyptians, including the significance of stars to them, is derived from the so-called Pyramid Texts. These hieroglyphic texts represent ritualistic utterances—spells, hymns, incantations, litanies, and glorifications—that aimed to protect and rejuvenate the dead pharaoh's body in the afterlife.[15] Moreover, they represent the oldest religious writings in the world, long predating the Old Testament scriptures. They were carved on the internal walls of five pyramids at Saqqara, the oldest pyramid being that of Unas, who was the last pharaoh of the Fifth Dynasty (about 4,300 BC). The other four were built in the Sixth Dynasty, the youngest of these having stood for about 4,100 years. The texts and the beliefs that the Pyramid Texts depict, however, are thought to derive from much earlier, predynastic concepts.

What these chambers in the pyramids of Saqqara make perfectly clear is just how important five-pointed stars were to the early Egyptians. The ceilings and corridors of many chambers are covered with dramatic constellations of these stars, emphasizing their importance in beliefs in an afterlife in the heavens. More than a thousand stars are crowded together on the chamber ceiling within the Unas pyramid, and such decorations became commonplace in many burial chambers. On the ceiling of the burial chamber of the pharaoh Merenre, the stars are packed so close together that their tips are almost touching. Similarly, row upon row of golden stars are spangled across the ceiling of the burial chamber of Amenophis II (approx. 1,410 BC).

Writing about Egyptian decorative art in 1895, Flinders Petrie observed that the natural ceiling pattern in Egyptian art was often of golden stars on a deep blue background, specifically a black night-blue rather than a dark daylight-blue.[16] Moreover, they were always five-pointed stars, frequently with a circular spot, usually red, in their center. When Tja-nefer looked closely at his fossil he would have seen not only the five-pointed stars on the top and bottom, but a little circle in the middle of each: a cluster of small plates called the apical system on the top, the mouth on the bottom. Is it any wonder that Tja-nefer would have been impressed by a domed stone found in the rocks, one which carried the selfsame symbol that adorned the ceilings of sacred chambers? One can only imagine how he thought the stars came to be on these fossils—did he think, perhaps,

Star hieroglyph from temple of Seti I, Luxor, Egypt. Photo by J. Bunbury.

that they were engraved there by another priest who lived long before he did, but who also was as fascinated by stars? Maybe they were sent from the heavens as a sign from those who lived an eternal life among the constellations. Could he have wondered, or even been aware of, the possibility that the five-rayed pattern on the fossil may even have been the very inspiration for the five-pointed star motif, long before his time?

Petrie made another observation about the early Egyptians and stars. He seems to have been one of the few people to have addressed the question of the association between the five-pointed pattern and the pinpricks of light that we see in the night sky. Petrie's reasoning was simple: to people of normal vision, the objects in the night sky appear as pinpricks of light—not five-pointed patterns; but to those of us who have gone through life nearsighted, the star in the sky is surrounded by a streamer of light. This, he suggests, indicates that "the Egyptians were short-sighted people from the early ages."[17] But why five streamers of light, he doesn't say. I suspect that the real explanation is a little less myopic.

Five-pointed stars are also found on the walls of the burial chambers in the pyramids at Saqqara. Yet rather than figuratively representing the constellations, they form part of the hieroglyphic Pyramid Texts. When it is used on its own as a hieroglyph, the five-pointed symbol can still represent a star. When used in combination with other symbols in hieroglyphic texts, it becomes what is known as a determinative. This is a symbol in hieroglyphic writing that confers a concept. In the case of the star symbol, it often forms, as I indicated earlier, part of words that are associated with time, such as *morning*, *month*, and *tomorrow*. The symbol also has connotations associated with priestly activities.[18] It is found in the hieroglyph for "a priest serving in a temple," "the priesthood," and "to adore." Sometimes, when it is used to depict the Duat—the starry afterworld of the god Osiris—it has a circle around it, representative, perhaps, of eternity, and making the symbol look even more like a fossil sea urchin. Another symbol in the word *duat* is a dome. This makes the sound "t," but it also, in its shape, mirrors the profile of Tja-nefer's fossil. The Pyramid Texts tell the fate of the deceased pharaoh:

> O King, you are the Great Star, the companion of Orion, who traverses the sky with Orion, who navigates the Duat with Osiris. (PT 882)[19]

The star symbol in hieroglyphs played a major role in their original deciphering. Before Jean-François Champollion began his long period of unraveling the mysteries of hieroglyphics in the early part of the nineteenth century, it had been thought that stars had been used to depict the actual positions of real stars in the sky. However, Champollion showed that often this wasn't the case. He argued that the star symbol was a "sign of the type," in other words, a determinative, signifying the nature of the figures or the group of hieroglyphs with which they are associated. As he wrote,

> The star of the inscriptions of Dendera is therefore the *last* hieroglyphic sign of each of them, and must be considered, not as the representation of a star, but as a simple element of the hieroglyphic writing; that is to say, as a kind of *letter*, and not as an imitation of an object.[20]

The recognition of this, supported by some other determinatives on the Rosetta Stone, was to be a major step forward in the elucidation of the meaning of hieroglyphs.

Napoleon Bonaparte was another who had been mistaken about what the star symbol represented in Egyptian hieroglyphics. Following his return from his Egyptian campaign in 1799, during which time the Rosetta Stone had been discovered, Napoleon decided to get rid of the fleur-de-lis symbol of the old French royal dynasties. He replaced it with the bee, which was the hieroglyph for Lower Egypt, and which the Romans had considered to signify the symbol of royalty in Egypt. With this the five-pointed star was sometimes also used in Napoleonic images to signify "divine king." At the time it was thought (quite incorrectly, as it turned out) that to the early Egyptians the star meant "divine."[21]

Much has been revealed about early Egyptian beliefs from finding the Pyramid Texts. Yet their discovery is not credited to one of the intrepid European Egyptologists who during the later part of the nineteenth century were crawling all over Egypt searching for hidden tombs and archaeological treasures to take back to their wealthy benefactors. Quite the opposite. The find, if the apocryphal story has any validity, is said to have been made by another type of scavenger—a jackal. The story goes that in 1879, eight years before Worthington Smith excavated Maud and her baby, a jackal had been seen by a workman at dawn near a crumbling pyramid in the necropolis of Saqqara. As he watched, the jackal wandered to the north face of the pyramid, briefly stopped, and then disappeared down a hole. Intrigued, the man decided to follow, wondering where and why the jackal had gone underground. Crawling first through a small hole and then down a narrow tunnel, he finally emerged into a chamber. And as he lifted his light the walls came alive with hieroglyphic inscriptions that cascaded from the top of the chamber to the bottom—inscriptions beautifully carved into the limestone walls of the pyramid, then painted over with turquoise and gold. What became of the jackal, history does not relate.

The texts themselves contain many allegories concerning stars, one of the central themes being the strongly held belief that the dead pharaoh would be reborn in the heavens as a star, and that his soul would travel up to the sky to find an eternal resting place in the stellar world of Osiris (Sah)-Orion, in the region known as the Duat.

The King is a star in the sky among the gods. (Utterance 586A, PT 1583)

My bones are iron and my limbs are the Imperishable Stars. I am a star which illuminates the sky. (Utterance 570, PT 1455)

This was the god of the dead, and of the resurrection. Thus Sah or Osiris was identified with Orion, while Isis (Sothis) was linked with the Dog Star, Sirius. During life the pharaoh was thought of as the reincarnation of Horus (Sopdu), the first man-king of Egypt, and son of Isis (Sothis) and Osiris (Sah). Sopdu, in whose quarry Tja-nefer found his fossil—the domed mound of rock engraved on either side with stars—was himself the product of gods linked with stars. In the Unas pyramid, there are many references to the dead pharaoh, Unas, as Sah-Unas. Moreover, they speak of the mummified Unas as becoming a star, in the constellation of Orion.

Crucial to the rebirth ritual was that the dead pharoah, as Sah, was reborn through the ritualistic act of mummification, performed by his sister-wife Sothis. With appropriate magical incantations pronounced over the mummified corpse, his soul would rise to join Sah in the constellation of Orion, as a star. Many images of Sah-Orion occur in ancient Egyptian drawings. One of the oldest is on the capstone of the pyramid of Amenhemet III. Here Sah (Osiris) is shown striding forth, carrying a large star in his hands. As the Pyramid Texts relate,

> Behold, he has come as Orion, behold, Osiris has become as Orion . . .
> O King, the sky conceives you with Orion, the dawn-light bears you with Orion . . . You will regularly ascend with Orion from the eastern region of the sky, you will regularly descend with Orion in the western region of the sky, your third is Sothis pure of thrones, and it is she who will guide you both on the goodly roads which are in the sky in the Field of Rushes. (Utterance 442, PT 820–22)

The role of Sothis, the mother of Sopdu, is central to the resurrection myth. The Dog Star, Sirius, to whom she is linked, descends below the horizon for about seventy days each year, and disappears into the Duat.[22] This period also marks the duration of the whole mummification ritual. The time when Sirius rises, or itself is reborn, in July, corresponds to the annual flood of the river Nile. Moreover, it marked the beginning of the ancient Egyptian calendar year. Because Sirius always rises immediately after the constellation of Orion, the two, Sirius (Sothis) and Orion (Sah), were always linked. Many of the statues of Sothis show her with a five-pointed star on her head. And as Sothis is reborn, so at the end of the mummification ritual the opening of the mouth ceremony was carried out, when the soul of the deceased was reborn and released to spend eternity among the heavenly stars.

There is a general consensus among those who have studied the Pyramid Texts that the rebirth ritual of the pharaoh depicted there was based on a reenactment of the story of Sah (Osiris) and Sothis (Isis), and the miraculous impregnation and birth of their son, Sopdu (Horus). From numerous references in the many funerary texts, including the Pyramid Texts, it is possible to reconstruct the Sah-Sothis-Sopdu story.[23]

Sah is the eldest son of Nut, the sky goddess, and Sothis is one of his sisters. He is seen as both a man as well as a god, becoming the first king of Egypt, with his sister Sothis as his consort. He is the finest of kings, establishing law and teaching religion and the arts of civilization. As a result, Egypt is prosperous and at peace. But there is one fly in the ointment—his brother Seth. First he plots against Sah; then he kills him and dismembers his body, scattering the pieces all over Egypt. Worse still, Sothis still has no children and there is no heir to Sah. However, not to be put off by her husband being murdered and cut up into many pieces, Sothis gathers the parts all up, and like an ancestral Dr. Frankenstein manages to reconstitute him by magical means, thus turning Sah into the first mummy. Now that he has come back to life, albeit for a short while, she has sex with him and becomes pregnant. Having fulfilled his manly duties, Sah transforms himself into a star, in Orion, becoming ruler of the Duat, the heavenly kingdom of the dead. Hiding from Seth, Sothis (Isis) gives birth to her son Sopdu (Horus) in the marshes of the Delta near Heliopolis.

> Your sister Isis comes to you rejoicing for love of you. You have placed her on your phallus and your seed issues in her, she being ready as Sothis, and Har-Sopd has come forth from you as Horus who is in Sothis . . . and he protects you in his name of Horus, the son who protects his father. (Utterance 366, PT 632–33)

When grown up, Sopdu challenges Seth to a duel, to fight for the right to be the ruler of Egypt and become Sah's true heir. During the fight Horus loses an eye, while Seth comes off worse, some might think, losing his testicles. Although there is no outright winner, Re-Atum, the creator of the Sun and the Earth, judges in Sopdu's favor, and he is proclaimed the first pharaoh.

Thus all subsequent pharaohs saw themselves as the reincarnation of Sopdu. When the pharaoh, as Sopdu, died, he knew that he would be reborn with Sah in the afterworld of the Duat. This offspring of the two stars was also seen as a star himself, and he was sometimes known as the "morning star":

> My sister is Sothis, my offspring is the Morning Star. (PT 357, 929, 935, 1707)

Tja-nefer's interest in the fossil urchin now becomes much clearer: the fossil found in the quarry of Sopdu—Sopdu the morning star, son of Sothis, the star Sirius, and Sah, the constellation Orion. It then becomes more likely that the quarry was so named because, like many limestones in this part of the world, it frequently contained fossil urchins. The lone specimen in the Museo Egizio is very unlikely to have been the only one taken by priests to Heliopolis. We can see Tja-nefer, visiting the quarry, being offered the fossil urchins that had been so assiduously collected. He then takes them back to Heliopolis, where they played, perhaps, an important role in the rituals associated with the rebirth of the pharaohs.

> There is another sacred bird named the Phoenix. I have never seen it myself, except in pictures, for it is extremely rare, only appearing, according to the people of Heliopolis, once in five hundred years, when it is seen after the death of its parent. If the pictures are accurate its size and appearance are as follows: its plumage is partly red and partly gold, while in shape and size it is very much like an eagle. They (the Heliopolitans) tell a story about this bird which I personally find incredible—the Phoenix is said to come from Arabia, carrying the parent bird encased in myrrh; it proceeds to the temple of the sun and there buries the body. In order to do this they say that it first forms a ball as big as it can carry, then, hollowing out the ball, it inserts its (dead) parent, subsequently covering over the aperture with fresh myrrh; the ball is then exactly the same weight as it was at first. The Phoenix bears this ball to Egypt, all encased as I have said, and deposits it in the temple of the sun. Such is their myth about this bird.[24]

So wrote the fifth-century Greek historian Herodotus of the legend of the phoenix. Since that time this legend has diversified and become embroidered to such an extent that to the Egyptologist Rundle Clark, writing in the late 1940s, the symbol of the bird rising from the ashes to signify rebirth and renewal has become emblematic of any appropriate resurrection theme, be it religious, political, or personal. The phoenix myth, he considered, "is one of the most evocative symbols ever devised by the human imagination."[25] Yet if we go back to the Egyptian texts in which it is described, such as the Pyramid Texts, and the later Coffin and Funerary texts that have survived, then from what is at times a contradictory and confusing story emerges an important religious belief. The phoenix "cult" was centered in Heliopolis, and it is clear from when it is first recorded in the Pyramid Texts that the phoenix story, even from these early times, forms part of the rebirth and renewal rituals involving the deceased pharaohs. As such it also has links with the stars and with Sopdu. And if we examine the details of these beliefs it becomes even clearer why Tja-nefer was so keen to take his fossil urchin back to Heliopolis with him, for I believe it might well have played a significant role in the rituals carried out by him and the other Heliopolitan priests.

When hieroglyphics were first being deciphered in the early nineteenth century by the likes of Jean-François Champollion in France and Thomas Young in England, it was noticed that a heronlike bird appeared to have played a central role in early Egyptian religious symbolism, even though it appeared not to represent a god as such. Known as Bn.w (usually referred to today as Benu), early Egyptologists believed that the phoenix was the Benu bird, or at least had been inspired by it.[26] One of the problems with this interpretation is the identity of the bird. Sometimes the Benu was clearly a heron or a lapwing. At other times it appeared to follow more closely the account of Herodotus and was a bird of prey, such as an eagle or a falcon. Subsequent research seems to indicate that the same symbolism may be attached to different representations of the bird: sometimes a heron, at other times an eagle. The Pyramid Texts talk of a relationship between the Benu and Sopdu. One predynastic god was a heron, and its identity seems to have merged with that of Sopdu in the Pyramid Texts.

One of the aspects of the phoenix myth is the bird's habit of perching on a mound. This mound is sometimes known as the pyramidion, or the *bnbn* (Benben) stone, the capstone of a pyramid.[27] Many Egyptologists have argued that this stone represents the primordial mound from which grew the lotus flower out of which emerged Re-Atum. However, this mound cannot also have been the resting place of the Benu bird. Nevertheless, the Benben stone played an important role in the religious ceremonies undertaken at Heliopolis, and much discussion has taken place in Egyptological literature over the last two centuries concerning the nature of this stone. In later depictions, the Benben stone is a pyramidion, but in the Pyramid Texts, which represent some of the older beliefs in ancient Egypt, the stone is drawn as a rough, conical shape with curved sides. This suggests that the pyramids at Giza were not simple scaled-up copies of the Benben stone, as some have believed. But what would Tja-nefer have made of his fossil? Not only was it inscribed top and bottom with stars, but in its conical shape it was probably very similar to the Benben stone—much smaller, maybe, but in all other respects a Benben stone.

And at Heliopolis what perched on the Benben stone was the phoenix or Benu bird. In early Egyptian beliefs, birds played an important role as go-betweens—flying to and from the known, living world and the unknown world of the dead. Birds had the power of flight and so could soar to the heavens. The primary and dominant incarnation of divine power in Egypt was the falcon, in the form of Horus (or as Sopdu). As Rundle Clark described it, this divine bird's "eyes were the sun and the moon and its flight was that of every heavenly body."[28] It was a bird of prey, and thus powerful—the very quintessence of omnipotence. Some of the earliest Egyptian religious beliefs saw other birds as representing the soul. The stork and the crested ibis were considered as playing the roles of the souls of the dead. In the Pyramid Texts the word *ba*, which is translated as "bird soul," is depicted as a stork. This bird soul reflected the power of the deceased pharaoh's soul to rise up into the world of the stars. It may even have been thought that the soul actually assumed the form of one of these birds. As such the bird was deified and became a mythological figure or a star.

Given the central roles played by the Benu bird, or phoenix, and the conical Benben stone on which it perched in Heliopolitan rituals; given the association of the phoenix with Sopdu, sometimes a bird, a falcon, like the phoenix, at other times the morning star; and given the central role of Sopdu, along with the celestial Sothis and Sah, in the whole rebirth ritual carried out at the death of a pharaoh, is it any wonder that Tja-nefer took such care to inscribe this star-etched, domed rock with his name, and the name of Sopdu? So maybe, as with the Neolithic skeletons of Whitehawk, Maud and her child, and the numerous other Neolithic, Bronze, and Iron Age bodies placed in northern European graves with fossil urchins for company, the pharaohs' journey to the afterworld may also be thought to have been facilitated by the judicious use of such strange little star-struck stones.

> Ouer & besides, I will not ouerpasse one kind of eggs besides which is in great name and request in France, and whereof the Greeke authors haue not written a word: and this is the serpents egg, which the Latins call Anguinum.
> PLINY THE ELDER, *Natural History* (1634, trans. Philemon Holland)

12

Snakes' Eggs

He was tall, with a fine head of white hair. Many considered him handsome. But—he limped. He drooled and slavered at the mouth. He stuttered. His nose dripped incessantly. His head shook and trembled. He was forever suffering illnesses of one sort or another. For much of his life he was ignored, mainly because he was an embarrassment to all around him—his family especially, who considered him an idiot. The Roman historian Suetonius reported that even his mother "often called him a monster: a man whom Mother Nature had begun work upon but then flung aside."[1] Yet despite all this he became emperor of Rome, ruling from AD 41 to AD 54, for he was Tiberius Claudius Drusus Nero Germanicus, third emperor of the Julio-Claudian dynasty and ruler of what was then the world's greatest empire.

Despite appearances to the contrary, Claudius was anything but a fool. Pliny the Elder, writing in his *Natural History*, a book that formed the basis for much of the European understanding of the natural world for well over a millennium, considered Claudius among the foremost scholars of the day. If he was anything, Claudius was an enigma. His accession to the lofty heights of ruler of the Roman Empire was every bit as bizarre as much of his behavior while he was emperor. His "election" as emperor was all over in a few hours. Caligula, Claudius's uncle and predecessor, had the dubious distinction of being the first Roman emperor to be murdered. Tradition has it that following his assassination by the Praetorian Guard, soldiers began looting the imperial palace. As they did, they came across a terrified Claudius, cowering behind a curtain. Carrying him off to their camp, the soldiers declared him emperor. Whether this military coronation was as impetuous as this or planned beforehand is a moot point. Either way, the senate had no option but to accede to the wishes of several thousand armed soldiers, who made it quite clear that they supported Claudius—and if anyone wanted to object they could take it up with some shiny, cold steel.

The reign of Claudius was one of the most eventful of any Roman emperor. He invaded Britain, following in the footsteps of Julius Caesar almost one hundred years earlier, and annexed much of the country. At home his reign was less glorious. His relationship with the senate was anything but smooth. Although he made attempts at conciliation with the leading council of Rome, his energetic and at times simply bizarre, unpredictable nature resulted in members feeling deeply resentful and bitter. Much to the horror of the essentially conservative incumbents of the senate, Claudius saw nothing wrong in appointing "long-haired" Gauls as senators, perhaps, in part, because he himself had been born in Gaul. Although he might have seen this as a demonstration of fairness and equality to all members of the empire, to the established Roman senators it was just another reflection of the irrational behavior of this fool who had been thrust upon them. During his reign some thirty-five senators and several hundred knights were either driven to suicide or executed.[2] Such behavior hardly made for a harmonious and properly functioning senate.

Scholarly and careful in many matters, Claudius was, at other times, almost psychotically cruel and savage (although he wasn't in the same sadistic league as Caligula). He loved gladiatorial "games." He was particularly keen on watching defeated opponents being executed. Moreover, even if one of his gladiators accidentally stumbled and fell to the ground, Claudius would have the errant fellow's throat cut. However, it was in the handling of judicial cases that he demonstrated his most unpredictable, and at times bloodthirsty, behavior. There is no doubt that he was diligent, almost to the point of being obsessive, in being present at court. But Suetonius noted that "sometimes he was wise and prudent, sometimes thoughtless and hasty, sometimes downright foolish and apparently out of his senses."[3]

In addition to the senatorial court, he instigated his own imperial court. Here he particularly liked to conduct treason cases. These courts could be held anywhere, even in Claudius's bedroom. And it was on these occasions that he made his most irrational and arbitrary decisions. While he would tolerate being shouted at and called a stupid old idiot, or even put up with, on one occasion, having a stone tablet and pen thrown at him so hard that it gashed his cheek,[4] he could also be most viciously vindictive. Once, after a man had been found guilty of forgery, someone in the crowd who had attended the trial shouted, "He ought to have his hands cut off."[5] Claudius duly obliged. He promptly sent for an executioner, with block and cleaver, to act on the suggestion. Another of his more celebrated instances of a savage ruling involved Valerius Asiaticus, a Gallic ex-consul and a former friend. However, having fallen from grace, he was convicted, seemingly on very little evidence, and forced to suicide there and then, in front of the emperor.[6] Such erratic behavior brought Claudius open and widespread contempt.

But it was in all probability a fossil sea urchin that influenced one of his most peculiar decisions—a decision that was not based on any particular feeling one way or another for the fossil. Instead, it was based on what this *ovum anguinum*, or "snake's egg," as this fossil was then known, stood for.

> Ouer & besides, I will not ouerpasse one kind of eggs besides which is in great name and request in France, and whereof the Greeke authors haue not written a word: and this is the serpents egg, which the Latins call Anguinum. For in Summer time yerely, you shall see an infinit number of snakes, gather round together into an hpape, entangled and enwrapped one within another so artificially, as I am not able to expresse the manner thereof: by the means therfore, of the froth or saliuation which they yeeld from their mouths, and the humour that commeth from their bodies, there is engendred the egg aforesaid. The priests of France called Druidae, are of opinion, and so they deliuer it, That these serpents when they haue thus engendred this egg do cast it vp on high into the aire, by the force of their hissing; which being obserued, there must be one ready to latch and receiue it in the fall again (before it touch the ground) within the lappet of a coat of arms or soldiours cassocks. They affirme also that the party who carrieth this egg away, had need to be wel mounted vpon a good horse and to ride away vpon the spur, for that the foresaid serpents will pursue him still, and neuer giue ouer vntil they meet with some great riuer between him and them, that may cut off and intercept their chase. They ad moreouer and say, that the onely marke to know this egg whether it be right or no, is this, That it will swim aloft aboue the water euen against the stream, yea though it were bound and enchased with a plate of gold. Ouer and besides, these Druidae (as all the sort of these magicians be passing cautelous and cunning to hide and couer

their deceitfull fallacies) do affirme, That there must be a certaine speciall time of the Moones age espied, when this businesse is to be gone about, as if (forsooth) it were in the power and disposition of man to cause the moon and the serpents to accord together in this operation of engendring the egg aforesaid by their froth and saliuation, I my selfe verily haue seen one of these egs, and to my remembrance, as big it was as an ordinary round apple: the shell thereof was of a certaine gristly and cartilagineous substance, and the same clasped all about (as it were) with many acetables or concauities representing those of the fish called a Pourcuttle, which shee hath about her legs. And it is the ensigne or badge that the Druidae doe carry for their armes. And they hold it a soueraigne thing, for to procure readie excesse vnto any princes, and to win their grace and fauour; as also to obtaine the vpper hand ouer an aduersarie in any sute and processe of law, if one do carrie it about him. But see how this vanitie and foolish persuasion hath possessed the minds of men! for I am able vpon mine owne knwledge to auouch, that the Emperor *Claudius Caesar* commanded a man of arms and gentleman of Rome, descended from the Vocantians, to be killed for no other reason in the whole world, but because he carried one of these egs in his bosome, at what time as he pleaded his cause before him in the court.[7]

If we are to accept the veracity of Pliny's account of the unfortunate death of the Vocontian soldier from Gaul, and there is no reason why we should not, it raises two significant questions. First, exactly what was this *ovum anguinum*, or snake's egg? And second, even bearing in mind Claudius's predilection for the homicidally dramatic, why kill a man for such a trifle? Although Pliny's description of the *ovum anguinum* is not particularly clear, medieval and renaissance naturalists were not dubious about its identity. Despite the discrepancies between a spherical object that "will swim aloft aboue the water euen against the stream" and a solid fossil made of rock that would just as likely float as fly of its own accord, some, like the great Swiss naturalist Conrad Gesner, were in no doubt that the "snake's egg" was a fossil sea urchin.

In his highly influential *De rerum fossilium, lapidum et gemmarum* (1565), Gesner depicted an example of what he accepted to be an *ovum anguinum*, viewed from above and below. The quality of his woodcuts is so good that his figure undoubtedly represents a fossil sea urchin. But rather than being an irregular urchin such as a shepherd's crown or a thunderstone (in other words, a heart urchin like *Micraster*, or the domed *Echinocorys* or *Conulus*), Gesner's fossil is a regular urchin, like those that the inhabitants of Santoigne in France regularly perforated. Doughnut-shaped and with

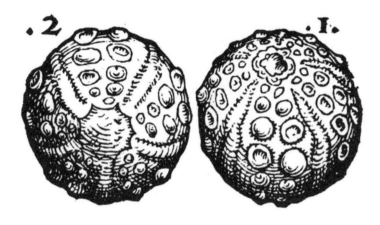

Conrad Gesner's 1565 woodcuts of an *ovum anguinum*, actually a fossil cidaroid urchin.

pronounced openings at either end (one for the animal's mouth, the other for its anus), his drawing is of a particular regular urchin called a cidaroid, although, as I noted in chapter 3, he failed to recognize it as such. These urchins are characteristically studded with blisterlike protuberances set in irregular rows. In life these were tubercles to which were attached long, stout spines that articulated in a ball and socket joint, and which the animals could wave around like a mighty array of protective, jousting lances. And insinuating themselves furtively between the tubercles wind five ambulacra.

Whether Gesner was merely articulating common folklore in his book or had himself been influenced by earlier writings is unclear. In all probability he would have been well aware of Pliny's account of *ovum anguinum*—how the stealer of the "egg" leaps high over a seething mass of serpents, tossing a fossil sea urchin (seemingly some sort of cidaroid) into the air and carefully catching it in a cloth. If dropped to the ground, its magical powers would instantaneously disappear. Could Mr. Wort's revelation to Herbert Toms in 1929 of the old habit of tossing a shepherd's crown three times in the air, then catching it (see chapter 8), be a distant echo of the practice recorded by Pliny two thousand years earlier?

Other naturalists supported Gesner's view that what today are regarded as fossil sea urchins were *ovum anguinum*. Michele Mercati, in his *Metallotheca Vaticana* (1574), likewise illustrated fossil sea urchins and called them *ovum anguinum*. Perhaps he didn't know any better and was just following Gesner's lead. Or then again, maybe he was influenced by the

A captured *ovum anguinum*, as depicted in what is said to be a 1497 woodcut.

prevailing view in France at the time that fossil cidaroid urchins were *les oeufs des serpents*. It is interesting that recently two fossil urchins, both species of the heart urchin *Micraster*, one collected in the nineteenth century, the other in the early twentieth, were recognized in the collections of the Muséum du Havre.[8] Each is accompanied by a label identifying it as a *coeur de serpent*, or "snake's heart."

In Wales these snakes' eggs or adderstones, as they were sometimes also known, were called *glain naidr*, while in Cornwall they were called *milprev* or *milpref*.[9] But why the supposition that they were snakes' eggs, and what was the significance of this? It is possible that the link between the druids and the snakes detailed by Pliny arose from the fact that the druids were also sometimes called *naddred*, or adders, alluding to the supposed regeneration, or rebirth, undergone by initiates as they similarly cast off the skin of the old life. If so, then the power attributed to the *ovum anguinum* mirrors the practices carried out in earlier Neolithic and Bronze Age times of placing fossil urchins in graves—a practice that may well have been tied up with the concept of resurrection of the spirit, in effect providing insurance that the deceased would be reborn. If indeed fossil urchins were used in this way in the pre-Celtic world, then they form part of an ancient belief system centered on the rebirth of the spirit in an afterworld.

Like Gesner, Mercati also distinguished irregular fossil urchins as something quite different from the *ovum anguinum*. He illustrated a form that is probably what we call *Galerites*, which is found in Cretaceous chalk deposits, and called it *Scolopendrites lapis*. This name was applied to this particular urchin because its five ambulacra were reminiscent of millipedes crawling over its surface, albeit glacially slowly.

Gesner, followed by Mercati, had further distinguished two other types of fossil urchins as *brontia* and *ombria*. His doing so emphasized the belief that these forms, rather than being serpent spawn, came from the heavens, and were the Roman equivalents of thunderstones. Moreover, this distinction perpetuated the view that Pliny had expounded in his *Natural History*, the source of many ideas on the natural world until Renaissance times:

> Ombria, which some call Notia, is said to fall from heauen in stormes, showers of raine, and lightening, after the manner of other stones, called thereupon Cerauniaand Brontia: and the like effects are attributed to it, as be reported of Brontia: and thus much moreouer, That so long as it lieth vpon the hearth of an altar, the libaments will not burne that be offered thereupon.[10]

Pliny's views on *brontia* and *ombria* were still accepted in the sixteenth century. Georg Agricola, in his influential *De Natura Fossilium* (1546), reported that "the ignorant believe that these [*brontia*] fall when it thunders. If it is raining when they fall they were called *ombria*." In Germany *brontia* were called Donnerkeile, while *ombria* were known as Regensteine.[11] Robert Plot, writing in *The Natural History of Oxfordshire* in 1677, referred to stones that

> (by the vulgar at least) are thought to be sent to us from *inferior Heaven*, to be generated in the *clouds*, and discharged thence in the time of *thunder* and violent *showers*: for which reason, and no other that we know of, the ancient *Naturalists* coined the suitable names, and called such as they were pleased to think fell in the *thunder Brontiae*; and those that fell in *showers*, by the name *Ombriae*: which though amongst other authors has been the only

reason why these have had place next the *stellated stones*, yet methinks it is due to most of them, by a much better pretence, having something upon them that rather resembles a *star* of five points, than anything coming from the *clouds* or the fish *Echinus*; to the shell whereof deprived of its prickles, *Ulysses Adrovandus*, and some others, have compared them, and therefore called them *Echinites*. However, I think fit rather to retain the old names, though but ill applied to the nature of the things than put myself to the trouble of inventing new ones.[12]

As recently as 1763, Richard Brookes[13] described how some people believed fossil urchins to be the eggs of tortoises because of the similarity between the plating structure revealed on a weathered fossil and the carapace of tortoises. Even today, the ideas of fossil urchins as *ovum anguinum* and resembling a tortoise are preserved in the scientific names of two of the most common species of the heart urchin *Micraster*. Found in the northern European chalk deposits, *Micraster coranguinum* and *Micraster cortestudinarium* have names that mean "little star with snakes" and "little star with tortoise," respectively.

While irregular urchins like *Micraster* and *Echinocardium* were known as *brontia* and *ombria*, it was regular urchins that were called *ovum anguinum*. Although Plot observed that such fossils were sometimes known as "*Porcupine-stone* without bristles,"[14] he noted how

> by Boetius and Gesner, and all the old *Authors*, they are called *Ova anguina*, Serpents eggs; perchance because from the basses there issue as it were five *tails*, of *serpents*, waved and attenuated toward the upper part of the stones. They tell us also a story of its being engendered from the *salivation* and *slime* of *snakes*, and cast into the Air by the force of the *fibulations*, where if taken, has effects as wonderful as its generation, and therefore of great esteem amongst the *French Druids*. But I care not to spend my time in *Romance*, and therefore proceed.[15]

If we are to accept that the *ovum anguinum* was, at least in many instances, a fossil sea urchin, then Pliny's account of it in his *Natural History* represents the first written record we have of an explanation of fossil sea urchins. According to Pliny, the Celtic druids believed that *ovum anguinum* was a powerful and magic object: it was an antidote against poison, as well as having the power to ward off various illnesses. The *ovum anguinum* was clearly a magical object that was also imbued with the power of good to ward off evil. If you needed success in a battle, why worry about armor or the sharpness of your sword—just place your faith in the trusty *ovum anguinum* tucked down your shirtfront. If your neighbor was getting a bit uppity about the fact that you had purloined a few of his sheep, you didn't need to spend half the day pleading your case—all you had to do was wave your *ovum anguinum* at him, and he would be bending over backward to stuff as many of his sheep onto your land as he could. And if a major dispute arose that called for adjudication by the emperor himself? Not a problem, thought our Vocontian knight. Right is might, and if it isn't, it's certainly an *ovum anguinum* stuffed down your bosom. The trouble with trusting a cold lump of stone, though, is that while you might be carried along with your belief in its powers, those not of a Celtic or druidic persuasion might not. Indeed, pleading that your entire case centered on your belief that this serpent's egg proved you were in the right and that it was your destiny to be victorious was, as it turned out, like waving the proverbial red rag in front of one of the more bullish emperors. The Vocontian lost more than his case—he lost his life.

The Romans despised the Celtic priests known as druids, and found many of their behaviors distinctly unpalatable. Despite being bloodthirsty themselves, particularly as revealed by their gore-filled gladiatorial contests, they were horrified by some of the druidic practices. A number of classical authors expressed their distaste for the druidic practice of human sacrifice. In his *Histories*, Diodorus Siculus wrote:

> When they attempt divination upon important matters they practice a strange and incredible custom, for they kill a man with a knife-stab in the region above the midriff, and after his fall they foretell the future by the convulsions of his limbs and the pouring of his blood, a form of divination in which they have full confidence, as it is of old tradition.[16]

Strabo in his *Geographies* recorded that

> the Romans put a stop to these customs, as well as to all those connected with the sacrifices and divinations that are opposed to our usages. They used to strike a human being, whom they had devoted to death, in the back with a sabre, and then divine from his death struggle, but they would not sacrifice without the Druids. We are told of still other kinds of human sacrifices; for example, they shoot victims to death with arrows, or impale them in temples, or having devised a colossus of straw and wood throw into it cattle and wild animals of all sorts and human beings, and then make a burnt offering of the whole thing.[17]

To Claudius, divining the future from the death throes of a human sacrifice was not only abhorrent, but a mockery of the Romans' own practice of sacrificing animals and reading the future in their entrails—the sacred ritual of *haruspices*. In Claudius's eyes this was a noble pursuit, and one that he much encouraged. These barbarian Celtic priests needed to be rooted out and destroyed. His dispatching the Vocontian knight in apparently so malicious a manner may well have been the act of an inherently coldhearted emperor prone to irrational and barbaric acts. But it may also have been a reflection of the concerted efforts made by a succession of Roman emperors to stamp out many of the Celtic sacrificial practices, and a general distaste for druidic belief systems.

Although many fossil urchins have been unearthed from Neolithic and Bronze Age burial sites, *ovum anguinum*, in the form of the pocked doughnutlike urchins, are not so frequently recorded from such archaeological sites. Maybe if they had been artificially changed by having a hole drilled through them, for example, then their chance of being spotted by some sharp-eyed archaeologist and so recorded for archaeological posterity would be much greater, as I will discuss below. The other reason for the relative rarity of *ovum anguinum* in Celtic and Roman times may be that it was an object more useful to the living than to the dead. Its value lay in its perceived power in everyday life, rather than in the afterlife. As a result, few have survived the ravages of archaeological time, though one *ovum anguinum* (a flint mold of a regular urchin called *Phymosoma*), set within a little bronze locket, is known. It was found in a Gallo-Roman temple near Essarts in the Seine Valley near Rouen, France.[18]

The practice of wearing *ovum anguinum* as amulets seems to have persisted into the early twentieth century in some parts of France. Writing on the survival of the tradition that regards these fossils as possessing magic powers, one Brother Pérot recorded how he met a woman on her way to the market at Clermont-Ferrand. Around her neck she wore a fossil cidaroid urchin, hanging from a piece of red wool. Pérot looked at her with astonishment. She stopped. He asked her what she was wearing around her neck. "If I told you," she replied somewhat tersely, "you would not understand, you town-dweller." She began to walk away, but Pérot persisted with quizzing her. "Is it your family's lucky charm?" "Exactly," she replied, turning around. "My old parents wore it, and I never leave it at home when I go to Clermont. It is not a thing to lose or to leave alone."[19]

From the fifth to eleventh centuries AD in northern Europe, other types of fossil urchins seem to have been considered special enough to be carried around with people, probably as lucky charms. A number of examples have been found in Merovingian (fifth to eighth century AD) and Viking (eighth to eleventh century AD) excavations in Denmark, northern Germany, Poland, and Russia of bound (or banded) pendants in which fossil sea urchins were encased in metal loops and used in all probability as amuletic pendants.[20] Such pendants were made from bronze, silver, or iron bands, or sometimes twisted wire. A typical example is a Viking-age fossil *Galerites*, caged within four bands of iron twisted together at the top, from Breininge on the island of Lolland in Denmark.[21] In addition to fossils, a variety of other objects were used in crafting the pieces, such as glass beads, cowrie shells, clay balls, amber, and even walnuts or fruit pits.[22]

Without exception when found in burials, the pendants have been associated only with adult women. Large pendants, like those containing fossil urchins, were worn tied at the waist.[23] In all likelihood they were perceived as protection against lightning, and against evil. They have also been found associated with cremations. In the late Iron Age site of Møllegaard in Funen, Denmark, the fragmentary remains of the metal bands from a pendant were found with a fossil sea urchin inside a vessel containing cremated human remains.[24] At this time unbound fossil urchins were still being placed in funerary urns. Two such examples were found in a cemetery at Westwanna in northern Germany.[25]

But what to do if you didn't have a fossil sea urchin to fashion into an amuletic bound pendant? The answer lies in the Statens Historiska Museum in Stockholm. Among the thousands of Viking items is a rather strange, small bronze brooch. Two animals, thought to be goats, flank what is generally considered to be a stylized fossil sea urchin in a bound pendant. Four raised, crenulated ridges arch across this conical central structure. They meet at the top of the dome in what is essentially a metallic knot.

The brooch was found at an archaeological site called Birka. This had been a trading place located on one of the islands in a lake called Mälaren. Here people lived and traded for two hundred years, between AD 750 and AD 970. The population has been estimated to have been between 700 and 1,000. Excavations have taken place several times on Birka. Beginning in 1871 the archaeologist Hjalmar Stolpe dug here for twenty years, excavating parts of the settlement and close to 1,100 graves of the estimated total number of 2,000.

Grave Bj973 was an inhumation burial, but the skeleton was not preserved. The grave goods indicate that a woman had been buried in that grave. In addition to the bronze brooch, the grave contained two oval bronze clasps, one round bronze clasp, one silver pendant in the form of a shield, forty-six beads (glass, silver- and gold-foliated, and carnelian), a pair

Bronze brooch from Viking grave Bj973 at Birka, Sweden, depicting Thor's two goats, "Gaptooth" and "Toothgnasher," flanking a stylized sea urchin in a bound pendant. Photo by Jan Lamm, courtesy of Statens historiska museum Stockholm.

of tweezers (iron and bronze), a pair of scissors of iron, a knife of iron, a whetstone, a needle case of iron, and one weight of iron.[26] Two additional graves on Birka have been found to contain the same type of bronze clasps featuring stylized fossil urchins. Both of these graves are female inhumation burials. All three graves have been dated to the latter half of the tenth century, during the Viking age. Another almost identical brooch of a similar age was found on the island of Bornholm in Denmark and is kept in the collections of the National Museum in Copenhagen.

As I have indicated, the bound pendants that used fossil urchins were tied with four metallic loops. In the bronze brooch we see an artistic representation of this very object, complete with its knotted metal clasp. If you were unable to find your own special protective thunderstone, what better than to make your own? And, moreover, one that is flanked by Thor's two goats, Toothgnasher and Gaptooth, reinforcing the close link between thunderstones and Thor, the god of thunder. Herein lies, I believe, the secret to why fossil urchins were so revered as amulets and why, possibly for thousands of years, they had been buried with the dead in many parts of northern Europe.

The prolonged interest in fossil sea urchins during the so-called Viking age was revealed in the ninth-century settlement of Haithabu, established near the present-day Schleswig in southern Jutland. For about 250 years, until it was destroyed by fire in 1066, it was the main commercial center for the region. Indeed, it has been argued that Haithabu was the first place in northern Europe in which people lived together in a citylike settlement. During excavations in the late 1970s within the semicircular bank of earth that surrounded the settlement, a total of 185 fossils were found.[27] Of these, more than two-thirds (128) were fossil sea urchins. Two forms were dominant: *Echinocorys*, represented by 77 specimens, and *Galerites*, another domed, helmet-shaped urchin, of which 31 were found. Both these fossil urchins are present in the moraines (glacial deposits) that are common in this area, having been derived from the chalk deposits by the scouring action of glaciers during the last Ice Age.

The fossils are preserved as flint molds, and many are eroded and look as though they were derived from beaches, supporting the view that people had deliberately brought them to the site. Moreover, one of the fossils shows signs of having been deliberately chipped. Mapping of the site showed an even distribution of the fossils. Their presence in excavations indicates that they were being collected throughout the period of occupation of Haithabu. In the view of Frauke Metzger-Krahé, the archaeologist who excavated the site, the concentration of the fossils was much greater than would be expected to have occurred by natural accumulation. Although most of the fossils were probably treated as thunderstones, Metzger-Krahé also suspected that some of the urchins had been used for practical purposes, possibly as playing pieces in board games.[28] For example, one urchin was found to have been polished smooth on its lower, flatter surface. During Viking times stones were often used in such games, examples having been found at Haithabu, and many are similar to fossil urchins in shape and size. In a Roman grave in Schleswig-Holstein a game board was found along with some smooth, rounded stone pieces, among which were found some polished sea urchins. Most of the Haithabu urchins, though, had probably once lined shelves or been placed by the doors of the settlement's cottages, to protect the houses from lightning or some other misfortune.

Although between one thousand and two thousand years ago *ovum anguinum* were thought to be of more use to the living than to the dead, the long tradition of burying fossil urchins such as *Micraster* and *Echinocorys* with the dead persisted well into Anglo-Saxon times, even after the coming of Christianity. For example, when an eighth-century Merovingian cemetery in the Pas-de-Calais area in France was excavated, a most unusual sight greeted the excavators. The graves were, for the most part, very simple. The skeletons were simply covered by a bed of paving stone. One grave, however, was slightly more elaborate than the others, consisting of slabs of stone laid in the form of a cross over a skeleton. As a Christian grave this was not very remarkable. What was surprising, though, was that decorative grave goods had been buried with the individual who occupied the grave. Placed on the head of the deceased was a large oyster shell, and along the sides of the body were some modern cowrie shells. And placed on each thigh was a fossil urchin flint *Micraster*.[29] Old habits certainly die hard.

While it has been generally thought that the ancient practice of placing grave goods with bodies had died out toward the end of the first millennium with the advent of Christian burials, this was not the case. Indeed, many other such examples of the continued placement of objects in Christian graves, from knives to keys and from bells to beads, have been excavated. Although there was a gradual diminution in this practice from the seventh to the eleventh centuries, graves such as the eighth-century ones in the Pas-de-Calais were not that unusual in this regard. As recently as the twelfth or thirteenth century, fossil urchins were still being buried with the dead. In a Jewish cemetery in Winchester, Hampshire, that was in use between 1177 and 1290, a child was buried with a fossil urchin that was placed by its head.[30] Like other objects placed in graves at this time, the urchin may have been thought of as a magic object that would play its role in protecting the body against evil forces or enhancing the child's chances of being resurrected.[31]

The earliest record of fossil sea urchins being found in an archaeological site is from an Anglo-Saxon graveyard at Chatham Lines in Kent, and goes back to the early days of the discipline of archaeology itself. In his book *Nenia Britannica; or, A Sepulchral History of Great Britain* (1793), the Reverend James Douglas (1753–1819), parson-antiquary, geologist, artist, novelist, and sometime mercenary soldier, described the discovery of some small bones. These had not been buried alone, but had been accompanied by a fossil sea urchin, along with eight amber and two variegated beads.

Douglas, who was chaplain to the Prince of Wales, was the first antiquary to provide detailed plans of a site that he had excavated. Eight years earlier he had read a paper to the Royal Society in which he argued that, in part on the basis of fossils, the undisturbed geological deposits in north Kent did not support the idea of a Universal Deluge.[32] He recognized that the fossilized animals and plants, though apparently of tropical origin, had lived where they were found. Thus the climate must have been appreciably warmer than in his day. Moreover, the forty-day flood recorded in the book of Genesis would have been of insufficient duration to transport animal remains from far away. Douglas believed that the Earth possessed an unknown power that could fossilize organic remains. Although others had proposed similar ideas independently, taken together they were of far-reaching intellectual significance.[33] Not surprisingly for the times, however, his views met with less than an enthusiastic response.

The fossil urchin that Douglas discovered in the grave was found at the side of the skeleton with one of the amber beads.[34] He was convinced that the fossil had been deliberately buried with the body:

> On the first discovery of this fossil, I considered it as having been thrown into the grave with the adventitious soil; but as it was a petrefaction [sic] of the *yellow silex*, and therefore heterogeneous to the black siliceous and white sparry echinites found in the chalk, its native bed having been of loam and gravel; and from its position also in the grave, among other relics deposited with the body, I am inclined to believe it was carried about the person when living, as an *amulet*.[35]

Not only was this the first recorded example of a fossil urchin from an archaeological site, it was also the first record of any type of fossil found in such a context—and the beginning of a realization that people had been collecting fossils for a very long time. Because the fossil urchin seemed to be so intimately associated with the body, Douglas considered it to have been a lucky charm that the person was strongly attached to in life—a personal *ovum anguinum*, perhaps. Was this the only reason why it had been buried with the person? Or could there have been a perception that the luck associated with the fossil may continue with the deceased into the afterlife? If so, was it the fossil as a whole entity or the five-pointed star alone that carried the power?

Following Douglas's discovery, fossil urchins have been found in a number of other Anglo-Saxon burials in England. Although some of these burials were probably pagan, others, as I have pointed out, may well have been Christian. One of the more striking was the grave of a woman from a fifth- to seventh-century Saxon-period site at Westgarth, Bury Saint Edmunds in Suffolk that contained a fossil *Echinocorys*. She wore around her neck a strand of colored glass beads; on her shoulder, bronze brooches; around her waist, fourteen blue glass beads; and tightly grasped in her right hand, a fossil sea urchin.[36]

Other skeletons have been found in which fossil urchins had been placed in very specific places in their graves. In third-century AD graves excavated in Frilford, Berkshire, in the 1920s, the fossils were found placed by the joints of the skeletons.[37] Similarly, a skeleton of a woman (whose skull had been artificially deformed) in a sixth-century AD grave in Stössen, Germany, was found with a fossil sea urchin placed carefully between her knees.[38] And a mid-sixth-century "rich woman's grave"[39] from Sarre in Kent contained a single specimen of the heart urchin *Micraster*. Fifteen hundred years after Maud died, people were still sending their spirits into the afterlife accompanied by lone fossil urchins.

The widespread practice, in parts of France, the Near East, and Saharan Africa, of drilling a hole through fossil urchins spread as far north as England. In an early seventh-century Anglo-Saxon grave at Buckland, Dover, a regular fossil urchin called *Cyphosoma* was found that shows signs of attempts to drill a hole through it.[40] A shallow depression, a couple of millimeters deep and about five millimeters wide, had been gouged into the surface of this hard flint urchin. Maybe it was the hardness of the flint fossil, but the driller seems to have been defeated and failed to complete the hole.

However, despite these examples, it would appear that from Iron Age times onward, when the Celtic culture was spreading across much of Europe, fewer fossil urchins appear to have been placed with burials, compared with Neolithic and Bronze Age times. Yet the fact that some urchins continued to be buried well into the eighth century suggests the possibility that pre-Celtic beliefs persisted, and were not entirely swamped by new ideas. More likely they were amalgamated with the new Celtic belief systems. Changing burial practices may have been a reason for this decline in the use of urchins as grave goods, with the increasing incidence of cremations at this time. But even here, as I have discussed earlier, fossil urchins were still sometimes placed with the ashes of the departed. The Tunbridge Wells *Micraster*, along with the stone axe in association with the cremation, is one such example. Yet the thunderstone link, seen by the fossil urchin's interment with a stone axe, is indicative of Norse beliefs, rather than the Celtic ideas derived from central Europe. By Anglo-Saxon times, the significance of fossil urchins was changing, from one of a deeply held spiritual belief that they would aid in the passage from this world into the next, to a more prosaic one that perceived them as magical objects that would protect against misfortune as well as bring good fortune. In life and in death, these little star-crossed stones were seen as being inextricably linked with their owners' fate—with their destiny. And to many peoples their destiny was linked to the stars, either in this life or in the next.

> His fate hovered near, unknowable but certain:
> it would soon claim his coffered soul,
> part life from limb. Before long
> the prince's spirit would spin free from his body.
>
> *Beowulf*, lines 2421–24

 # 13

Star of Destiny

A core concept that has permeated Indo-European myth for at least six thousand years is that of an all-powerful Fate or Destiny. The brooding strength of Fate was ever present and ran "like a scarlet thread through the tales both of the gods and the great heroes."[1] H. R. Ellis Davidson succinctly observed further:

> The men of the north knew that they walked along a precipice edge, their precarious security threatened constantly by the sword, the storm, the attack of an enemy. Disaster might reach them at any time, whether heralded by gloomy omens or falling from a blue sky. The crops might fail, the fish disappear . . . from the sea . . . A blunted sword-blade, a chance encounter, or a snow-storm in the hills might result in the loss of all that made life worth living, and even the most far-seeing could not guard against such accidents.[2]

In Old English this Fate was called Wyrd,[3] "a dread omnipotent personality to whom even the gods were subject."[4] It was the lot of Wyrd to spin the thread of human life for all eternity. A deep-rooted origin to this belief is shown by the many European cultures—Celtic, Germanic, Norse,

Roman, and Greek—in which Wyrd takes the form of three female deities controlling the fates of men. Even gods and goddesses were not exempt from the brooding shroud of Fate, for they, like humanity, were also subject to its pervasive power—and in the thrall of the weavers of the destiny of the gods, and of humanity.

In Anglo-Saxon England the three women together were known as the Wyrd. During the Renaissance they manifest as the three witches in Shakespeare's *Macbeth*. In Old Norse only one of them was called Wyrd (or Urdr); the other two were Verdandi and Skuld, and collectively they were known as the Nornir. To the Romans they were the Parcae, the three being named Nona, Decima, and Morta. To the ancient Greeks they were known as the Moerae.[5] Always dressed in white, their names were Clotho, Lachesis, and Atropos—and they were the daughters of Night.[6]

What linked this female trinity across all these cultures was that their own fixed destinies were to spin and weave the threads of fate for all humanity—one spun the thread, another measured its length, and the third applied the fatal cut. Of the three Moerae, Atropos was the smallest, but she was considered the most terrible. The thread of life was spun on Clotho's spindle; it was measured by the rod of Lachesis; and it was cut by the shears of Atropos.[7] To some, Zeus, the mighty god, was thought to weigh the lives of men and inform the Fates of his decision. But others held that even Zeus himself was subject to the power of the Fates. The Moerae were not his children but born of the Great Goddess of Necessity "against whom not even the gods contend and who is called 'The Strong Fate.'"[8]

Such myths were widespread in the eastern Mediterranean, Robert Graves noting their importation into Greece from Crete, Egypt, Palestine, Babylonia, and elsewhere.[9] The concept of the three Fates has a long Indo-European heritage extending back at least six thousand years.[10] Because of their white robes the three Fates were considered to be Moon-goddesses. The word *Moera* literally means "a share" or "a phase," and the three Fates were equated with the three phases of the Moon. The Great Goddess of Necessity who begat the Fates was immortal, changeless, omnipotent. In art she was often depicted with a lime-whitened mound (under which the harvest corn doll was hidden)—or mounds of seashells, or quartz, or white marble.[11] And when they were constructed, the barrows that silhouette the chalk ridge skylines today would have been stark, white mounds, within which it was the fate of lone kings—and sometimes their fossil sea urchins—to be buried.

In the Icelandic *Poetic Edda*, under the guise of the Nornir, the three Fates are depicted as appearing at the birth of Helgi, the future king:

> Then was Helgi, the huge-hearted
> Born in Bralund to Borghild.
> Night had fallen when the Norns came,
> Those who appoint a prince's days:
> His fate, they foretold, was fame among men,
> To be thought the best of brave kings.
> There in Bralund's broad courts
> They spun the threads of his special destiny:
> They stretched out strings of gold,
> Fastened them under the hall of the moon.[12]

Life's destiny was set out at birth as interwoven threads of gold that stretched across the starry night sky. The threads woven by the Fates unfold during an individual's life. The inexorable link between weaving and destiny is shown by the Old English saying *me þæt wyrd gewæf*, or "fate wove me that destiny." The Old English word *gewæf* means "wove"; its cognate, *gewif*, means "fortune."[13] Individuals' threads of destiny were interwoven with those of others to produce "life's rich tapestry," with each life linked with others in a complex web of fate that could at times cast a kindly smile—or at other times cast a malevolent glare that bespoke disaster. There was no rhyme or reason to the fickleness of fate. Only the shamans, the druids, or those who had access to the thoughts and motives of the Wyrd could ever hope to explain the past, justify the present, or foretell the future.

When the radical new religion of Christianity began to replace "pagan" religions in northern Europe, people struggled to embrace one practice and give up the other entirely. Whereas it eventually became transmuted into "Providence and the will of Almighty God,"[14] Fate remained the driving force in the great Anglo-Saxon poem *Beowulf*. To the writer, Fate was omnipotent. As Seamus Heaney points out in his admirable translation of *Beowulf*, Wyrd spreads a claustrophobic and doom-laden atmosphere over all the characters in the poem; they "all conceive of themselves as hooped within the great wheel of necessity."[15] Throughout the poem, Fate hovers close by, "unknowable but certain."[16] At its culmination, Beowulf goes to meet his fate, entering a barrow to fight a dragon who in turn,

> Swaddling in flames, . . . came gliding and flexing
> and racing towards its fate.[17]

Despite having accounted for the monster Grendel and his mother, Beowulf meets his end in the dragon's barrow, uttering his final words: "Fate swept us away."[18]

For much of the Anglo-Saxon period, neither Fate nor Christianity gained the ascendancy—Wyrd was all-powerful. Even the Christian God was subject to her powers, for, in the words of the Anglo-Saxon poem "The Wanderer," "Wyrd's will changes the world." Many other surviving Anglo-Saxon poems from the seventh and eighth centuries stress the power of Wyrd. Its power could, at times, be destructive, and was to be feared and revered by all. This was conveyed in "The Ruin," a poem written in the seventh century on two leaves, now badly scarred by fire, describing a deserted Roman city. The poem begins:

> Well-wrought this wall. Wyrd broke it.

and

> Bright were the buildings, halls where springs ran,
> high, horngabled, much throng-noise;
> these many meadhalls men filled
> with loud cheerfullness: Wyrd changed that.[19]

And in "The Wanderer," the Maker of men is cast in the role of a Destroyer as inexorable as Wyrd itself.[20] The Wanderer reflects on past glories, although

> There stands in the stead of staunch thanes
> a towering wall wrought with worm-shapes;
> the earls are oft-taken by the arch-spear's point
> —that thirsty weapon. Their wyrd is glorious.[21]

There was no escaping Wyrd, who exerted her viselike grip on the troubles of humanity, for "Fate is restless."[22] Wyrd held sway not only over life, but also death, for it was the natural order of things. "A weary heart cannot oppose its inexorable fate."[23] It is the way things happen:

> And all is hardship
> On earth, the immutable decree of fate
> Alters the world which lies beneath the heavens.[24]

But by the ninth century things had changed for the most part, and God had gained the upper hand. The Fates had had their day. King Alfred the Great, writing in about the year 888, observed in his translation of Boethius's *De Consolatione Philosophiae* that *Ac þæt þæt we Wyrd hataþ, þæt biþ Godes weorc þæt he ælce dæg wyrceþ*: "What we call Wyrd is really the work of God about which he is busy every day."[25]

In some ways it should come as no real surprise that in the belief systems of many societies, destiny seems to have been thought of as being interwoven with spinning and weaving. The reason is that both these activities formed an integral part of life for most people for thousands of years. The sheep that had been domesticated by early Neolithic peoples yielded the wool. After being shorn the wool was washed and combed, then turned into yarn by being drawn into a thread by hand. This thread of yarn was then attached to a weighted stick, known as a spindle; the weight was some form of small, flattened stone disk known as the whorl. The thread of yarn was then spun once more, providing extra strength to the yarn.

The spindle whorl carried out quite a simple, yet crucial job—it transformed a thin, flimsy thread that could easily be broken into tough yarn that would last much longer. The whorls acted, in effect, as little flywheels on the spindle, stabilizing and prolonging the turning as it worked its twisting magic. Moreover, the spindle whorl prolonged the life of the rotating spindle. And the Nornir, the Parcae, and the Moirai spun the destinies of humans, twisting them into a thread that was sometimes tough and long lasting, but other times as thin as gossamer, and as transient as the passing shadow of a cloud.

The concept of the spindle was central to Plato's views of the structure of the universe and humanity's place in it, all heavenly bodies having been thought to have rotated about the "Spindle of Necessity." In his *Republic*, he recounted the "Myth of Er," describing how Er, who has been killed in battle, comes back to life rather fortuitously as he lies on his funeral pyre. He then tells of how his spirit traveled for days to wonderful, strange places. At one he and other souls could see a shaft of light stretching through Earth and to heaven like a pillar.[26] And from the end of the shaft of light hangs the spindle of Necessity, around which the heavenly bodies revolve on a series of spindle whorls.

> And the whole spindle turns in the lap of Necessity. And on the top of each circle stands a siren, which is carried round with it and utters a note of con-

stant pitch . . . And round about at equal distances sit three other figures, each on a throne, the three Fates, daughters of Necessity, Lachesis, Clotho, and Atropos; their robes are white and their heads garlanded, and they sing to the sirens' music, Lachesis of things past, Clotho of things present, Atropos of things to come.[27]

In *Timaeus* Plato talked of how the souls of men are distributed among the stars, one soul to each star, and being set in the stars "as it were in chariots."

And he who should live well for his due span of time should journey back to the habitation of his consort star and there live a happy and congenial life.[28]

In northern Europe this "pagan" significance of stars was hinted at by Bishop Wulfstan, Archbishop of York from 1002 to 1023. Despite more than half a millennium of being "Christian," a significant proportion of the population at the time still clung to old "pagan" ideas, including the belief in a connection between celestial and earthly events. Wulfstan railed against these people, some of whom, he said, believed "that the shining stars were gods and worshipped them earnestly."[29]

During the Renaissance, the power of the celestial was ineluctable. To Shakespeare's King Lear the stars were the ministers of Fate.[30] As Lear cries out at the end of his tragedy:

> It is the stars
> The stars above us, govern our conditions.[31]

In his *History of the World*, Sir Walter Raleigh wrote of the stars as asserting "complete power over all the reasonless things in the inferior world."[32] As such, they were seen as having a strong influence on the disposition of humanity and the nature of life. (Not that things have changed much. Even today the stars rule the lives of many who earnestly devour their horoscopes in their daily newspaper.)

In much Elizabethan and Jacobean literature in England, the stars are praised (to the heavens, presumably) or else blamed for their iniquitous influence on the mortal affairs of mankind—Romeo and Juliet, the quintessential star-cross'd lovers, are a case in point. Don Allen pointed out how this Renaissance philosophy is admirably summed up in Michael Drayton's poem "Endimion and Phoebe":

And that our fleshly frayle complections,
Of Elementall natures grounded be,
With which our dispositions most agree,
Some of the fire and ayre participate,
And some of watry and of earthy state,
As hote and moyste, with chilly cold and dry,
And unto these the other contrary;
And by their influence powerfull on the earth,
Predominant in mans fraile mortall bearth,
And that our lives effects and fortunes are,
As is that happy or unlucky Starre,
Which reigning in our frayle nativitie
Seales up the secrets of our destinie,
With friendly Plannets in conjunction set,
Or els with other meerely opposet.

The influence of the stars on the fortunes of all in society, from the king to the peasants, could be either benign or malevolent. When the stars looked kindly on people and were the bestowers of gifts of good fortune, they were called "blessed stars," "lucky stars," "happy stars," "smiling stars," or "fair stars."[33] But life was rarely a bed of roses during these times, and the cosmos took its fair share of the blame—the stars were accused of being malevolent, malicious, wrathful, and oppressive. They were "malignant and ill-boding stars," "cruel stars," or "unlucky stars,"[34] even sometimes the direct cause of pestilence. Just take a frequently used word, like *disaster* (which the early nineteenth-century paleontologist Louis Agassiz also applied to a fossil sea urchin—who says scientists don't have a sense of humor?). This word, which has its root in the Latin word *aster*, meaning "star," has a less than amiable connotation. It derives from the Old French word *disaster*, meaning "evil star"—something that goes against the stars, against our destiny. If our destiny is linked to the stars, then an evil star will cause misfortune to our destiny. While there were those who argued strongly against these ancient pagan beliefs, to the man and woman in the street, their fate and fortune might as well be blamed on the stars as on any more worldly cause.

What, then, is the link between fossil sea urchins and destiny in this life? Why did the Vocontian knight think that his precious *ovum anguinum* would bring him the luck he needed to win his case? Did he believe his destiny was entwined with that little star-crossed stone? Although the habit of burying fossil sea urchins with our dead may have been kept to ensure a destiny in the afterlife, one widespread use of fossil urchins, through both time and space, may provide a clue to the long association between the stars, fossil urchins, and destiny in this life. And this, as I mentioned in chapter 10 and describe above, was the use of these little fossils for thousands of years as spindle whorls. Here was a small stone the perfect size to place on the spindle. But more than that, it was a stone that came with its own imprinted star. This association could hardly have been missed by ancient peoples whose life was so tied in with ideas of destiny, and the power of the stars to influence it. A whorl—which they thought probably came from the heavens bringing its own star, and which prolonged the spinning of the thread—must have carried great spiritual resonance, over and above its basic utilitarian significance. And such thread woven into fabric must also weave into it the destiny ordained by the stars.

From late Paleolithic times right through to Roman and Saxon times, a period of more than thirty thousand years, fossil urchins were artificially perforated. Before the advent of weaving they were probably worn as ornamental beads, but not long afterward they were almost certainly used as spindle whorls. Determining whether a perforated urchin had been used as a bead or a spindle whorl comes down largely to its size, its weight, and the diameter of its hole. Any round, centrally pierced object greater than about eight-tenths of an inch (2 cm) in diameter is more likely to have been used as a whorl than a bead. Most have holes 7–8 millimeters in diameter, though these may be as narrow as 3–4 millimeters.[35]

In earlier chapters I have discussed how perforated fossil urchins have turned up in many archaeological sites, especially in France. In their exhaustive overview, François Demnard and Didier Néraudeau recorded 184 such urchins, from fifteen departments in seven regions, mostly in central, southwestern, and northern France.[36] Most fossils came from deposits in Charente and Charente-Maritime, and ranged in age from the late Paleolithic to Roman times, though most have been found in Neolithic to Iron Age deposits. As I have discussed in chapter 10, such drilled urchins have also been found in some Neolithic and Iron Age sites in Jordan. Clearly, this was a widespread activity, in terms of both space and time. The urchins most frequently perforated were regular urchins from Cretaceous deposits. With a natural opening on the top for the anus and one on the other side for the mouth, they readily lent themselves to having holes drilled through them. Support for the notion that these fossils once spun on a spindle is provided by the discovery, in at least five of these French deposits, of unmistakable spindle whorls made from baked clay and similar in size, weight, and shape to the fossils.

Given that it was probably much easier to make a spindle whorl from clay, the question has to be asked, why did these people go to so much trouble to laboriously drill a hole through hard stone? Unless the hole was drilled exactly in the center, it wouldn't function efficiently. Because late Palaeolithic urchins with holes drilled through them were unlikely to have been used as spindle whorls, their use was more likely to have been ornamental, or perhaps ritual. Indeed, it may well have been a ritualistic reason for why fossil urchins came to be used as spindle whorls in the first place. If this is the case, then the expenditure of the energy to produce a perfect, smooth hole became justified. As Roberta Gilchrist has pointed out,[37] spindle whorls may have been regarded as keepsakes or as protective amulets symbolic of the home. But they are also significant for their link with spinning and weaving, activities that sometimes carried undercurrents of magic. Gilchrist points to the argument of Burchard of Worms in his *Collector*, written about 1010, where he suggested a link between spells, charms, and the weaving of cloth:

> Have you been present at or consented to the vanities which women practise in their woollen work, in their weaving, who when they begin their weaving, hope to be able to bring it about that with incantations and with their own actions that the threads of the warp and the woof become so intertwined that unless [someone] makes use of these other diabolical counter-incantations he will perish totally? If you have ever been present or consented you must do penance for thirty days on bread and water.[38]

Further support for this argument comes from the discovery of spindle whorls that look for all the world like fossil urchins, but which were in fact manufactured to mimic the fossils. A spindle whorl found at Woodhouse, Northumberland, originally thought to be of Roman age but now considered to be no older than Tudor,[39] looks remarkably like a fossil cidaroid urchin, complete with five grooves to represent the ambulacra and raised, nipplelike structures resembling the tubercles on the fossil originals.[40] The only difference is that it was made from lead. Kenneth Oakley reported the discovery of a pottery spindle whorl from a Bronze Age site between 4,300 and 4,000 years old in Cyprus and now kept in the Limassol Archaeological Museum; it, too, was manufactured to mimic a fossil urchin.[41] If finding the real thing proved too difficult, then why not make a copy? What is more, attempts to drill a hole through an actual urchin met with failure, and what remained was a fossil with indentations on both sides. Yet these fossils were not always just cast aside. Two of the four fossil urchins found at the shrine at Wadi ath-Thamad in Jordan are just such an example.[42] Despite being of no apparent practical use in spinning, they must still have been imbued with some degree of spiritual significance.

Only one artificially perforated fossil urchin has ever been found in Fairford, England. In 1844 and 1845, thirty-six graves, subsequently found to be of Anglo-Saxon age, were uncovered.[43] Some of these finds were acquired by the antiquarian and numismatist John Yonge Akerman and mostly published and illustrated in his book *Remains of Pagan Saxondom* (1855). Of the more inconsequential finds from one of the graves that didn't make it into this publication was a single fossil sea urchin, subsequently described by Akerman as "A *vorticellum*, or Spindle-whirl [sic], formed of the lower portion of the flint cast of an echinus (*Galerites Albogalerus*)."[44] Despite more extensive excavations of these graves by William Wylie,[45] no more fossil urchins are recorded as having been found. However, an object acquired by Akerman,[46] and further examples found by Wylie, provide a link between the fossil and the symbolic pattern on its surface: the five-pointed star. The Fairford graves are renowned for having produced a large number of saucer brooches. These are large, dished brooches, worn by women in pairs on the shoulder. And the central motif carried by some of them is a five-pointed star.

The power of the star-crossed urchin to control man's fate and destiny became increasingly transferred from these naturally occurring objects, picked up from muddy fields or sun-baked hillsides, to the potent symbol itself, freed from the fossil's surface—the five-pointed star.

Faust:
The pentagram torments you?
Oh tell me, son of hell,
If it bounds you here, how did you get in?
How was such a ghost betrayed?

Mephistopheles:
Inspect it properly! It is not well drawn:
That one angle, which faces outward,
Is, as you see, a little open.
GOETHE, *Faust*

14
Five-Star Attraction

It was a typical November evening in Cambridge. The mist had set in mid-afternoon, and the sickly yellow street lights seemingly battled to make much headway against the myriad tiny droplets of water that hung in the air. I wandered along King's Parade, past the newly erected Christmas decorations—shimmering waterfalls of light, and the ubiquitous constellations of bright, shining stars—five-pointed stars, of course. I had come to Cambridge to give a lecture on evolution. But I couldn't miss the opportunity to hear some of the best choral music in the world.

A lone bell tolled through the mist, a muffled beckoning to Evensong in St. John's College Chapel in the University of Cambridge. It was decidedly drier inside. I made my way to one of the hard, oak pews that lined the sides of the chapel. The organ soon sounded, the choir processed up the aisle and bowed to the altar in the east. Scores were lifted, mouths were opened, and the first note of Gibbons's Magnificat—"My soul doth magnify the Lord . . ."—soared effortlessly toward the beautiful vaulted ceiling. I followed the sound, with my eyes as well as with my ears, up past the great organ pipes—and up to a single five-pointed star perched high above the longest pipe.

It wasn't the first five-pointed star I had seen that day. In addition to the sparkling Christmas decorations lining the street outside, I had seen hundreds emblazoned on small, round pebbles—fossil sea urchins nestling in drawers, like eggs in a carton, in the university's Sedgwick Museum. It struck me that here was a rather extraordinary link between an organ that is a few hundred years old, Christmas decorations, and some fossils that are tens of millions of years old—all carried prominent five-pointed stars. When the fossil urchins had been alive, the five-rayed star represented the external expression of the animals' internal plumbing system, a feature unique to all echinoderms—urchins, sea stars, brittle stars—that allows them to breathe, feed, move, and sense their environment. When stripped in death of the bristly covering of spines, the five-rayed pattern becomes very obvious. Almost but not quite as obvious as the rather surprising star clinging to the top of the organ.

I began to wonder whether I had been thinking too much about fossil urchins and five-pointed stars. Writing books can become a little bit obsessive—it seemed that I was beginning to see stars everywhere. But it wasn't my imagination. There it sat, as if to mock me, almost goading me into explaining just why it clung so tenaciously to its lofty musical seat. Perhaps it had been set there to represent the star in the east, one of our potent symbols for Christmas—the Epiphany Star, the celebration of Jesus's birth—the star that appeared to the Wise Men as a guiding light. Was this, I wondered, why we typically have such a positive feeling toward this pentagonal shape, and why it is becoming more and more associated with Christmas with each passing year? Of all the symbols we see around us today, it seems to be the one that offers a sense of hope, optimism, inspiration, and goodness.

Yet there is also a dark side to five-pointed stars. Think black magic, devil worship, and people prancing naked around a slaughtered goat, and you will probably also think pentagram—a five-pointed star. Why, then, this strange polarity in attitudes toward this symbol, whose written history extends back even to the very birth of writing, five thousand years ago?

One clue to the positive way in which five-pointed stars have long been regarded is found in the fourteenth century long poem *Sir Gawain and the Green Knight*.[1] This anonymous work reveals how in medieval times the five-pointed star, referred to as the Pentangle, featured in aspects of Christian symbolism that had nothing to do with the birth of Jesus. In the early part of the poem Sir Gawain prepares to leave for the castle of the Green Knight, setting off to challenge the knight on behalf of King Arthur:

Then they showed him the shield that was of shining gold,
With the Pentangle portrayed in pure gold hues.
He takes it by the baldric, and arranges it about his neck
That suited the knight seemly fair.
And why the pentangle pertained to that noble prince,
I am determined to tell you, though it should delay me.
It is a symbol which Solomon once set
In betokening truth, by title that it received,
For it is a figure that has five points,
And each line interlaced and locked in the other;
And everywhere it is endless, and the English call it,
Everywhere, as I hear, "the Endless Knot."
Therefore it accorded to the knight and to his shining armour,
For, ever faithful in five times in each way.
Gawain was known for his goodness, and as pure as gold,
Devoid of all villainy, with virtue displayed
 Without dispute.
 Therefore the new Pentangle
 He bears on his shield and tunic,
 As man of tale most true
 And most noble knight of speech.
First he was found faultless in his five senses.
And again, never failed the knight in his five deeds,
And all his trust in the world was in the five wounds
That Christ received on the Cross, as the Creed tells us.
And wheresoever this one in battle found himself,
His proud thought was in that through all other things,
All his courage he found in the five joys
That the gracious Queen of Heaven had of her Child.
For this reason the knight beautifully had
In the inside of his shield her image painted,
That queen he looked thereto his courage never failed.
The fifth five that I find that the knight used
Was Generosity and Fellowship beyond all things;
His Purity and his Courtesy, which were never corrupted;
And Compassion that passed all things. These pure five
Were bestowed more on that knight than on any other.
Now all these five things were forsooth settled on this knight,
And each one embraced the other, that had no end,

>   Being fixed upon five points that had no end,
>   Nor together never in any direction, nor ever broken,
>   Without end at any angle I anywhere can find,
>   Wherever the design began or came to an end.
>   Therefore on his bright shield was the knot
>   Royally with red gold upon heraldic red.
>   That is the pure Pentangle, so-called by wise people.
>   Now Gawain was ready and gay;
>   And taking his lance
>   He gave them all good day
>   He went, for evermore.[2]

So maybe that was it—the five-pointed star (or pentagram, or pentangle, or sometimes the pentalpha, literally meaning "five times the letter *A*")[3] was a token of the five senses, our five fingers, the five wounds of Christ, the Virgin Mary's five joys, and the five virtues of generosity, lovingkindness, purity, courtesy, and compassion. Add to this mix truth, faithfulness, and gentleness of behavior, and the five-pointed star appears to be imbued with more fine virtues than a saint. No hint of devil worship here. And it would seem that in some quarters these heroic and virtuous connotations of a five-pointed star are still with us, having emerged recently in a rather unlikely setting.

In Russia the five-pointed star may be set to make a comeback after falling from grace. As the symbol of socialism, the star featured prominently on the flag of the former Soviet Union, along with the sickle. When the union became the Russian Federation, however, the star disappeared. But late in 2002 Defence Minister Sergei Ivanov argued that it should be incorporated in the flag of the new, post-Soviet Russia: "Servicemen hold the star as something sacred (as) our fathers and grandfathers fought under the star." Today the five-pointed star sits proudly atop the Spasskaya Tower of the Kremlin. When first used in the early years of the Russian Revolution, according to the Central Museum of the Soviet Army, "It [meant] that the Red Army fights so that the star of justice should shine for the reaper-peasant and smith-worker . . . The five-pointed star is the emblem of the proletariat of the five continents of the Earth."[4]

After the choir had processed out of the chapel and the congregation had drifted off into the misty night, I lingered, gazing up at the five-pointed star. I couldn't help but feel that its presence on these organ pipes was a little incongruous, not only because of its place on the Soviet Union flag and on Sir Gawain's shield, but because of its traditionally ambivalent relationship with Christianity. But why a *five*-pointed star—why not four points, or six, or seven? Perhaps it had something to do with the idea that the first Christians interpreted the five points of the pentagram as representative of either the five wounds of Christ on the cross or the doctrine of the Trinity plus the two natures of Christ.[5] Indeed, early Greek Christians sometimes used it in place of the cross.[6] The five-pointed star has also been thought to signify "our Lord's Epiphany," the "revelation of the Christ Child to the Gentile wisemen," and was sometimes referred to as the Star of Jesse or of Jacob, reflecting heavenly wisdom.[7] Because of the association of the symbol with truth and the works of God, the Roman emperor Constantine I, upon his conversion to Christianity, included the pentagram in both his seal and his amulet.[8]

However, the use of the five-pointed star as a Christian symbol declined to medieval times, and its presence in some cathedrals and churches may owe more to the interests of the masons who built them than to any deep religious symbolism. Even as far back in time as the Roman Republic, the pentagram was used to symbolize the building trades, particularly stonemasonry.[9] Medieval masons regarded the pentagram as being symbolic of deep wisdom, which probably explains its presence in a number of cathedrals from this period throughout Europe.[10] For instance, the sketchbook of the thirteenth-century stonemason Villard de Honnecourt includes a number of drawings of pentagrams. In more recent times, the use of the five-pointed star on certain Masonic Grand Lodge seals and banners is probably an echo of this medieval use by stonemasons. It might also explain the prominent pentagrams that figure in the decoration of St. Paul's Cathedral in Melbourne, Australia, built in the late nineteenth century.

One argument for the prevalence of the five-pointed rather than a four-, six-, or seven-pointed star in modern cultural consciousness is its symbolic representation of the perfect human form. Earlier I alluded to why Neolithic people in Jordan might have viewed the five-pointed star on fossil urchins in this light (see chapter 10). Evidence that it was the five-rayed star that attracted these early fossil collectors is supported by the discovery in the 1930s of earthenware pots, dating probably from the fourth and fifth centuries BC, in Israel, at Azekah (Tel Zakariya) and Gezer, and in Jerusalem.[11] Some of these pots have engravings of five-pointed stars, in the form of pentagrams lying within a circle—really quite reasonable, figurative depictions of fossil urchins, like *Coenholectypus*. Another example is known from a frieze in the third-century-AD synagogue of Capernaum.[12]

Whether the influence was Semitic or Hellenistic-Greek is not clear. A tomb at Marissa whose Greek inscriptions indicate an age of about 200 BC has graffiti on it, including many pentagrams and hexagrams.[13] There may be a Pythagorean influence here. Undoubted use of pentagrams occur on a Jewish tomb in Tortosa, Spain, dated at about the fourth century AD.[14]

On an ancient Greek amphora, the pentagram appears on a warrior's shield. It was also used on many Greek coins, dating from the fifth century BC, for about seven hundred years. In a Pythagorean context the pentagram was the "Salus Pythagorae," a token of health and well-being. In a study of pentagrams, Jan Schouten has argued that the original source of that symbol was the Middle East, where it was a magic emblem associated with protection of the individual against evil or misfortune.[15] The Pythagorean symbolism, on the other hand, centered more on the pentagram's association with health and the harmony of the body. Even if, as seems probable, the later Greek symbolism was derived from the more ancient apotropaic symbolism, what was the source of the symbol in the first place, and how did it attain its attribute of providing protective powers? Could the answer lie in the many fossil urchins scattered across these lands, and which people have been collecting so assiduously for thousands of years?

And to the followers of Pythagoras in sixth-century-BC Greece, the five-pointed star, in the form of a pentalpha, personified balance and perfection, and was symbolic of the healthy individual.[16] The most famous pentalphic representation of the human form is Leonardo da Vinci's Vitruvian Man. In all probability a self-portrait of da Vinci, the head and outspread limbs in this drawing make a five-pointed star and are confined by a circle. Uniting the points by encircling them united the body with the spirit. Similarly, in Agrippa von Nettesheim's *De Occulta Philosophia*, published in 1531, and in Tycho Brahe's *Calendarium Naturale Magicum Perpetuum*, published half a century later, are drawings of a pentagram on which is superimposed a human figure.[17] As the points of the stars in Agrippa's illustration are signs of the zodiac, of Mars, Venus, Mercury, Saturn, and Jupiter, humanity is seen as being in harmony with the elements and with the universe.

Whether the five-pointed star perched above the organ pipes in St. John's College Chapel has Christian overtones, or had been put there by an avid stonemason, or had been placed in its heavenly position to perpetuate a Pythagorean tradition is debatable. I subsequently discovered that there is, in fact, a much more prosaic, and melodic, reason for its being there. The star is actually an integral part of the workings of the organ, and is known as a cymbelstern (literally, "cymbal star"). They are not uncommon on organs and are generally mounted near the top of the organ case. When activated the star spins, and small bells attached at each of its points strike a stationary clapper. But it still begs the question of why the cymbelstern is five-pointed, rather than four-, six-, or seven-pointed.

The five-pointed star, when not perched on a fossil urchin, had been used iconographically long before the birth of either Christianity or the ancient Greeks. To the Mesopotamians more than five thousand years ago, this star was symbolic of a heavenly body.[18] A thousand years later its meaning had changed. Known as UB, it meant "region," "Heavenly quarter," or "direction."[19] The symbol appears on clay tablets that bear protocuneiform inscriptions dating from 3000 BC to 2900 BC, found at a site called Jemdet Nasr, about sixty-two miles (100 km) south of Baghdad in Iraq.[20] The stars have also been found engraved on Proto-Elamite clay tablets (3000–2500 BC) as well as on spindle whorls and a cylinder seal, and painted and incised on pottery.[21] In ancient Egyptian tombs more than four thousand years old, five-pointed stars adorn the ceilings of burial chambers (see chapter 11). One is embossed on an Assyrian clay tablet used as loan tag more than 2,500 years ago, recording the loan of grain.[22] To early Hebrews (although they prefer the six-pointed Star of David today), the five-pointed star was the Seal of Solomon and represented the Pentateuch, the first five books of the Bible, and thus the concept of truth.[23] In northern Europe it was an ancient pagan (that is to say, pre-Christian) symbol believed to have the power of warding off evil. Consequently, in medieval times it was placed on a range of everyday items, such as bedsteads and cradles, as well as on building exteriors, the doors of cowsheds and churches, and the gates of castles.[24]

Although the practice of burying fossil urchins with the dead declined in the Middle Ages, people would have continued to place the fossils on windowsills or near doors to keep the devil and other evil forces out of the house, or build them into the design of their churches for the same reason. We know that people placed them in houses up to two thousand years ago because of the fossils found in the first- to fourth-century dwellings in Studland, and their common occurrence at the ninth-century town of Haithabu. But how to protect the house from the devil if no fossil urchin was to hand? Simple. Just draw the key feature—the five-pointed star—as a pentagram. This is what Faust does to try and keep the devil, in the form of Mephistopheles, out of his house. He fails, though, because, as Mephistopheles points out to Faust, the pentagram wasn't drawn properly, one of the points being left open, so it lost its power.

However, properly drawn pentagrams were thought to perform the same apotropaic function as fossil urchins. A German cradle dating from 1579 and held in the collection of the Bayerisches Nationalmuseum in Munich has a pentagram drawn on its interior to keep the devil away from the young child at night.[25] During the Middle Ages even adults thought it was sometimes advisable to have this insurance. A votive tablet from Stadeleck, Germany, dating from 1675 and also in the collections of the Bayerisches Nationalmuseum, features a painting of an elderly couple in bed, and on the foot of the bed is inscribed a large pentagram. During this period pentagrams were also drawn on the walls of buildings, on the gates of castles (such as at Bentheim), on the doors of cowsheds, and, mirroring the urchins set into the wall of St. Peter's Church in Linkenholt, at the entrances to church buildings. One example is in the porch of a Norman church at Knauthain near Leipzig, where a pentagram decorates the capital on the right-hand side of the entrance.[26] The church tower of the Marktkirche in Hanover, which was built in 1350, is topped by triangular gables ornamented with inset circles. Within one of the circles is a pentagram, set there, it has been suggested, to keep evil at bay.[27]

The practice of placing pentagrams or five-pointed stars on buildings for apotropaic reasons appears to have survived in the Fenlands of eastern England until the late nineteenth century, and it closely mirrors the emplacement of fossil urchins by doors and on windowsills that persisted well into the twentieth century. A Victorian cottage on the banks of the river Cam, built in 1897, would seem to be protected by prominent five-pointed stars. These stars, set within a circle and looking for all the world like fossil urchins, are carved into sandstone lintels on the cottage exterior: one above the front door and one each above the two windows. There are echoes here also of the emplacement of fossil urchins in Romano-British cottages at Studland nearly two thousand years earlier.

What makes it likely that the carved stars were apotropaic is that they were not placed centrally in the broad sandstone lintels, as might be expected if they were purely decorative. No, they had been carved at the very base of the lintels, so that they carefully touch the tops of the windows and the top of the door—as close as possible to the entrances of the dwelling, seemingly to protect the inhabitants of the cottage as effectively as possible against malevolent spirits.

To modern adherents of "pagan" spirituality (the Wiccans), each point of the pentagram (often drawn as a pentalpha—the endless knot) is said to represent an elemental force. The top left is Earth; the top right, Air; the bottom left, Fire; the bottom right, Water. The top point represents the elements of the spirit. There is no evidence that this symbolism existed in pre-Christian times. The modern interpretation derives more from Aristotelian concepts—the quintessence theory of the four elements and the spirit. But with a renewed interest in "pagan" concepts in the nineteenth century, the five-pointed star, when inverted, became the manifestation of evil. The pentagram shown "right side up," with the single point at the top, is a positive symbol, representing the "spirit over the body." However, turn it upside down and the black dogs of hell are unleashed, symbolizing, at least from the nineteenth century onward, evil and the devil.

On a more prosaic level, though, today we are surrounded by five-pointed stars. They splatter the plastic bags in which we take home our groceries. One sits on a packet of butter I opened at breakfast. Another stared at me from the fuel pump when I filled my car with petrol on the way to work. Coffee shops, beer bottles, soccer balls, you name it—the more you look, the more you come to realize that five-pointed stars really are everywhere. They appear on 63 national flags of the world, none more so than the U.S. flag, which contains 50. What's more, 16 U.S. state flags also bear stars. When the European Union decided on its flag, a circle of 12 gold five-pointed stars was chosen because the emblem was said to represent solidarity and harmony between the peoples of Europe. No reason was given for why five-pointed rather than, say, six-pointed stars were chosen. In one of the EU's member states, Italy, its one Euro coin has a circle of five-pointed stars around its periphery, and features Leonardo da Vinci's Vitruvian Man, arms and legs spread wide to make his own five-rayed pattern, in its center.

Yet one place I really didn't expect to find a five-pointed star was high in a remote part of the Derbyshire Peak district. Arguably, the most significant Neolithic stone circle in northern England is Arbor Low (said to be from the Old English *eorthburgh hlaw*, meaning "earthwork mound"). Perched high in limestone country in the southern Pennines in Derbyshire, its eighty-three-yard-wide (75-m), near-circular bank, up to thirteen feet (4 m) high and with an inner ditch, encloses a spectacular circle of forty-two huge limestone slabs. It is unclear whether these slabs—up to ten feet (3 m) long and arranged as if by a giant hand, with long axes pointing to the center of the circle—may once have been upright. Cemented to one, perhaps within the last two decades, is a plaster plaque, painted purple, within which had been carefully engraved a five-pointed star in the form of a pentalpha. Another, albeit unwittingly created, link between Neolithic sites and five-pointed stars.

The confusion in the minds of twenty-first-century Christians as to just what a five-pointed star symbolizes surfaced in England early in 2002. A parishioner in the village of Stradbroke in Norfolk petitioned the archdeacon of his local diocese for the removal of the star that perched on top of the village church tower.[28] Seemingly oblivious to the star's Christian connotations, he perceived it simply as a symbol of witchcraft. To the Stradbroke church's rector, though, the star brought joy and happiness to many people in the area. Certainly the positive aspects of five-pointed stars seem to outweigh the negative ones: if you have done well, you are a star (and at school when I was young you were literally given one—the gold ones being the most sought after). To many, our astrological star sign determines who we are, and how we behave—it represents our destiny.

But what of the humble fossil sea urchin that has carried this symbol of our fate and destiny on its back for so many hundreds of thousands of years? It can hardly have found a place in the new religion. But then again, in some places old habits die hard . . .

Nowhere is the link between the old and new religions better manifested than in two windows, each mantled by a crown of fossil sea urchins, at St. Peter's Church in Linkenholt (see chapter 1). The incorporation of the fossil urchins into the very fabric of the church in the thirteenth century, and the continuation of the motif when the church was rebuilt in the late nineteenth century, signifies the culmination of millennia of fascination with these fossils. One possible reason for these pagan symbols having been set above the window in the first place is its location on the north side of the church. This side was known in medieval and earlier times as "the side of the church where the Devil was supposed to lurk."[29] In later Saxon churches doors were placed in the north and south walls, near the western end of the building. The door on the north side was known as the Devil's Door.[30] It was thought that this door was where, during the early period of the Christian church, those who still clung to pagan beliefs could enter, for they wanted to continue to worship at the old pagan sites that the Christians had built over. The north door may have been used either as a means of deliberately segregating these heathens from the Christian believers, or perhaps as a way for pagans to find one another.

The identification of this door with the devil relates to the link that the early Christian church made between evil and the old pagan ideas that they wished to suppress. Another important use of the Devil's Door is said to have occurred at baptisms. During the ceremony the north door was left open so that any spirit of the devil lurking within the baby would depart through this door, which was quickly closed to prevent his coming back

Nineteenth-century window in St. Peter's Church, Linkenholt, Hampshire, decorated by twenty-five flint fossil urchins. Author photo.

into the church and reentering the baby. Given the widespread belief in the apotropaic powers of fossil urchins, it would not be totally unexpected to find them placed near such a door. Was there also a link here with the old Danish idea that fossil urchins protected the unchristened child from being "changed"?

We do not know for certain if the Linkenholt church ever possessed a Devil's Door, but it was in widespread use in early medieval churches. In the twelfth-century church at nearby Ashmansworth, it is possible to see the outline of a door on the church's north side that long ago had been bricked over. Such was the usual fate of these doors. For instance, of the thirty-nine Saxon churches documented in Sussex that possessed both north and south doors, seventeen have had their north doors walled up. This fate befell only one south door.[31] Perhaps if the medieval window and fossils at Linkenholt were once set next to a Devil's Door, they were seen as an added insurance in protecting the church against evil.

Whatever it was that prompted the rebuilders of St. Peter's Church in the late nineteenth century to continue the practice of placing fossil urchins above the window on the north side of the church—and, additionally, place them above the window on the south side and around a window on the nearby church school—it is unlikely to have been a desire to keep the devil out of the buildings. Most likely the objects had become highly ingrained in the culture of the village. It would be easy to think that the presence of fossil urchins in the Linkenholt church was a mere aberration—the product of some local fancy with these odd, but nonetheless entrancing, objects. But if you thought that, you'd be wrong.

Further to the south in Hampshire, a little to the east of Winchester, is a beautiful little wood where more fossil urchins quietly lie. The mature beech, oak, and rowan trees that grow there give the visitor a feeling that the place has been undisturbed for a very long time. Driving along the little road that passes by this patch of untouched woodland, it would be very easy to miss the little church that is set deep in the wood. It is known, quite logically, as the Church in the Wood (or more formally as the Upper Itchen Benefice Church in the Woods, Bramdean Common). This church, though, is not surrounded by a large churchyard filled with graves; nor is it the focal point of a village (indeed, it lies far from any settlement); nor does an obvious track lead to it. But if you look closely, a path of golden oak and beech leaves winds through the sparse undergrowth, and in the distance it is possible to catch a glimpse of a most unusual church.

Colloquially it is known as a "tin tabernacle," for the simple reason that it is made entirely from sheets of corrugated iron. Such churches were quite common in the latter part of the nineteenth century. Unlike the centuries needed to build some cathedrals, here was a very cheap and extremely quick way to build a church. In 1883 this one took a mere five days to erect. It was built "in order that the commoners, charcoal burners and gypsy itinerants who used Bramdean Common were able to attend a church."[32] In other words it would appear that they were not welcome in the nearby parish church. However, Church in the Wood is now managed and maintained by the rector of the local parish church.

The small churchyard that surrounds the church is demarcated by a wooden fence, and entry to it is through a small, white wicket gate. When I tracked down the church late one wet November day, it appeared to be

Church in the Wood, a Gypsy "tin tabernacle," Bramdean, Hampshire. The date it was built, 1883, has been formed using fossil urchins. Author photo.

floating in a sea of sodden oak and beech leaves that thickly carpeted the ground. It actually wasn't Church in the Wood that I had come to look at, but the ground immediately in front of the gate that was set a short distance from the church. The leaf litter there must have been close to a foot thick, and had obviously been accumulating for many years. Once I had brushed it away a mold-blackened step appeared, on which was studded some numbers—1883, the year the church had been built. Further brushing revealed brown tiles surrounding blackened concrete into which the studs were set. I set to work cleaning the studs, using the paleontologist's main tool of the trade for cleaning specimens—a trusty toothbrush. As I worked in the increasingly heavy rain, the studs slowly began to change color, from black to white. And there, faintly discernible on some of the better preserved studs, were five-pointed stars: set into the concrete 119 years earlier were 40 fossil sea urchins, all the helmet-shaped *Echinocorys*: 5 for making the one, 12 for each of the eights, and 11 for the three. So here, too, perhaps, were echoes of an ancient tradition and an adherence to a belief in the apotropaic properties of the fossils, placed by the entrance to the churchyard.

There can be no more ancient symbol of the power of good over evil than the five-pointed star, whether it be drawn on the wall of a house or on a cradle, or incised on a flint sea urchin and placed on a windowsill. Those who buried Maud and her baby some four thousand years ago felt so strongly about the symbol's power that they sent mother and child on to the afterlife with hundreds of fossil urchins. Maybe they thought that the fossils would protect the pair from any evil spirits that they might encounter during their long and perilous journey. Or perhaps the star emblazoned on each was thought to have even greater powers—the power to help the deceased achieve immortality, even to reach the starlit heavens. Wasn't this, after all, why Tja-nefer in Heliopolis was so interested in his fossil urchin, as a means of helping the pharaoh reach the stars in safety? Maybe this is why we often represent the pinpricks of light that spangle the night sky as five-pointed stars, not tiny circles, because of this distinctive emblem incised so powerfully on the fossil urchins. With this symbol has lain not only our destiny in this life, but also the key to rebirth in the afterlife.

More than twenty thousand years ago, depicting animals and the hunt in paintings in places like Lascaux in France and Altamira in Spain was in essence a spiritual expression as much as it was an outpouring of artistic creativity. By appearing as paintings on the walls of caves, the spirits of the wild animals were being created in the belly of the Earth, to be reborn into the light of day. Similar arguments have been made for the large Neolithic and Bronze Age burial mounds in northern Europe—by placing not only bodies but also objects of great spiritual importance in these womblike structures within the Earth, rebirth was being sought. The role of fossil urchins in such belief systems is most acutely demonstrated by the activity of Neolithic and Bronze Age inhabitants of Brittany. Why go to the trouble of building a great mound and placing just one fossil sea urchin at its center unless the fossil held some underlying great spiritual significance?

If the concept of fossil sea urchins as thunderstones extended back to Maud's time, perhaps she was taking them with her to return them to the Thunder God. By doing so, she may have been trying to ensure that she and her child would be reborn among the stars. In the process the thunderstones would be recycled to this heavenly deity, this protector of his earthly flock, this celestial shepherd who holds the stars in his hand. Like people for thousands of years before and after her, Maud may have thought that her destiny, and that of her child, lay in the stars, in life as in death. Over time this association between fossil urchins, with their striking five-rayed pattern, and a celestial rebirth evolved into a merging of the dual symbolisms of the five-rayed pattern and the stars in the night sky—the five-rayed pattern became a star, and the star became a five-rayed pattern. Perhaps this outcome was inevitable. Or maybe the concept spread from the East. While Maud was being buried with her cosmic coterie of protective thunderstones, Tja-nefer and his fellow priests in Egypt were already representing the stars, their dead pharaohs' celestial destination, as five-pointed patterns, as if lifted directly from the fossil urchins that these holy men collected—fossils obtained from the quarry of the god who in one guise was himself the morning star.

In an unexpected way, Maud's hope of finding a renewal for herself and her child was realized. Although her spirit and that of her child may have gone on their celestial journey, their mortal remains and her fossil treasures saw the light of day once again, when they were unearthed some four thousand years after they had been buried. For that she has an alert farmer and an enthusiastic amateur archaeologist to thank.

Walls of flint separate the world of cold and dark from that of warmth and light. Shortly after the last glimmer of sunlight fades into the west, pallid moonlight spreads itself across the shepherds' crowns that lie cold and still in the pot near the front door of the small house, the ghostly five-pointed stars offering a long-forgotten protection to the home. The year has almost ended and through the window can be seen a tree, symbolic in recent times of Christmas and of the birth of the Messiah. Its green branches bow under the weight of colored baubles, candles of light, and dangling presents. And standing proudly on the top, a symbol of kindness, of continence, of courtesy, and of compassion; a symbol of the harmony of the human form with the cosmos; a symbol that first fired the imagination of our ancestral species hundreds of thousands of years ago; and a symbol of good over evil—a five-pointed star.

> I'll sing you five-o
> Green grow the rushes-o.
> What is your five-o?
> Five for the symbols at your door.
>
> From "Green Grow the Rushes-O" (traditional)

EPILOGUE

The Marlborough Times and Wilts and Berks County Paper for October 6, 1871, carries a detailed account of the dedication of the newly rebuilt church in Linkenholt. In it are the names of many in attendance, from Hillary, the builder and William White, the architect, to the Reverend George Festing and the bishop. Marquees, large, open-sided tents, have been set up in the courtyard for the celebratory festivities. Hundreds mill around, greeting one another and admiring their beautiful new church. It is much better than the dilapidated old ruin that, as the bishop pointed out, had lived all its life among the pigsties and deserved to be put to rest after seven hundred years.

What words were spoken, I wonder, about the strange, round flints that formed a border around some of the church windows? Sitting in the churchyard 132 years later, I can imagine the swirling crowds dressed in their Sunday best, and the wind blowing around the lichen-covered gravestones now covering part of the churchyard—graves within which probably lie many of those who attended the celebrations that day, happy that life in the village was at last back to normal . . .

"Good morning, Mr. Hillary. Nice to see you here on this auspicious occasion."

"A fine mornin' indeed, Mr. White. Just the day to launch a church."

White smiles. "I see you did it, then."

"Why, did you think we wouldn't?" Hillary replies.

"I really wasn't sure if you and your brother would go through with it."

"Well, they're hidden away on the north side, just like they were on the old church, so I don't sees any harm'll come of it."

"I'm sure you're right, Hillary."

"In fact," continues the builder, leaning forward slightly and looking White straight in the eye, "after we'd popped 'em all in around the window, I stood back and looked at 'em and thought to meself, 'they look so good, p'raps we should have some more.'"

"What do you mean?" asks White slowly, with a horrible feeling rising slowly from somewhere beneath his ribcage.

"Haven't you taken a look at that other window right by the main door?" Hillary cocks his head toward the corner of the church where a cluster of people are at that very moment peering up at the window.

"Oh no," he sighs. "What have you done?"

But White doesn't need an answer. He had known Hillary and his brother from old, having employed them to build some of the other churches he had designed in the area. He isn't sure now whether he'd ever be asked to design another church again—he, William White, church architect of Wimpole Street, London. Even so, when he turns and sees what Hillary had done, he can't help laughing. "Well, Hillary, you and I will either suffer the fires of eternal damnation for that little trick, or we'll have the wrath of the bishop come down on our heads like a ton of, well, flints, I suppose!"

"Not sure which'd be worse," laughs Hillary.

"Does he know?"

"He gave me a friendly nod just now, so I don't think so. But looking at that gaggle over there pointing up at 'em, I think he soon will. Oh, and before you go, Mr. White, don't forget to check out the little school."

Fifty yards to the west, looking like a miniature version of the church, stands the one-room school that Hillary and his men have also built to White's specifications. It even has its own tiny steeple with a bell, to call the students to their lessons.

"Yes, I was just admiring how well it had turned out," says White. "Very fine indeed."

"Oh, fine it is, Mr. White," Hillary replies with a mischievous grin. "But what I meant was, go and take a good look above the window next to the front door."

White's mouth falls open. "Oh, Hillary, you didn't?"

With another grin and a tip of his hat, Hillary wanders off to the tent in search of a beer.

White can't help but be amused as he watches him go. Slowly shaking his head, he wonders how on earth he is going to explain to the bishop why his rebuilt church and brand-new school are festooned with pagan symbols—balls of flint, each etched with a five-pointed star—and dozens of them.

The consecration ceremony begins at noon, when the bishop and his entourage arrive: no less than thirty clergy joining the more than one hundred locals who aren't going to miss this event for the world. The clergy include the Reverend George Festing, who the year before bought the advowson of the rectory. Until then it had been with the manor, next to which it had stood since at least the early fourteenth century. So many people have come that they can't all fit in the church. Because of this, halfway through the consecration ceremony the bishop leads his clergy, along with those villagers who were able to get inside, out into the churchyard. He consecrates it, and then preaches his sermon from a small rise to the east of the church—"And Moses went into the midst of the cloud, and gat him up into the mount . . ." Then he and the crowd repair to a large tent containing tables heavy with food and drink.

Later, when the tables are substantially lighter, the bishop rises to his feet.

"I don't wish as this time to drink a toast, though I do wish to congratulate the incoming rector on the results of the day. It has been one of those singularly happy occasions which seems to have had prosperity from beginning to end. I must congratulate the landowner, Mr. Colson, and those families who had contributed to the rebuilding of the church. They have pulled down an old, worn-out church which had lived all its life among pigsties, and built a beautiful new church on a mountaintop."[1]

As the afternoon is fast disappearing, White decides that it is time to have a close look at Hillary's workmanship, especially those windows. While many people are now wandering from the tent and heading over to inspect their new church, none seem to have ventured around to the back. White thinks he will pop round for a quick look before anyone else does. But someone has beaten him to it. A young woman is standing beneath the window, looking up at it with what, to his relief, seems to be some satisfaction.

"It's quite unusual, isn't it," ventures White.

"To you, maybe, Mr. . . . ?"

"White. William White, the architect."

"How do you do, Mr. White?" Miss Tipley says, offering her hand and introducing herself. "You've done a wonderful job." She places her white hand on her hat as she looks up at the window. "Yes, this window might seem strange to you, Mr. White, but to those of us who have lived all our lives in the village, it is most reassuring to see the shepherds' crowns still here."

They both stare silently at the thin little window. The fossil urchins have been inset around three sides of it, just as they had been in the old church: ten fossils along the top, five running down on either side, and a final one placed horizontally next to the last on each side: twenty-two five-pointed stars protecting the church.

"Have you any idea why these fossils were built into the old church in the first place, Miss Tipley?"

"I can't say for certain," she says, and pauses. "But I have heard tell that it has something to do with the window's being next to the Devil's Door. The door was there either to keep the devil out or to expel him from church. I can never remember which, I fear. I shouldn't be at all surprised if the shepherds' crowns had been put there to make sure that the devil couldn't get in."

"Why do you say that?"

"Well, most of us keep shepherds' crowns by our front doors. My mother told me they used to say that it would stop the devil getting in the house. It always made me feel safer when I was a child. Silly, really. They're just old fossils. But I really must congratulate you for putting some more by the main doorway."

"Oh, yes, that. I promise you it wasn't my idea. We have Mr. Hillary to thank. I understand some of his workers from these parts put him up to it."

"I shall certainly feel very safe in church on Sundays now, Mr. White," Miss Tipley says with a smile.

Taking his leave of Miss Tipley, White returns to the front of the church to look at the other window. Immediately to the left of the old medieval doorway on the southwest side, now bathed in late afternoon sunlight, a window almost replicating the original has been installed. In keeping with the other "modern" windows, though, it has a trefoiled top instead of the more simple, rounded top of its medieval counterpart. The other difference is that in keeping with the arched design of all the other new windows in the church, the window's border of fossil urchins has been set in an arched pattern, crowning it. White counts twenty-five fossils in all. The one in the center of the arch is placed differently than all the others—its flat surface is showing, but it still has the five-rayed star. This is an urchin called *Conulus*, the selfsame specimen that the Swanscombe Man had been so attracted to four hundred thousand year ago. On either side of it were placed twelve urchins. All, save a single specimen of *Micraster*, are *Echinocorys*, and all look like carefully arranged eggs—but eggs with five-pointed stars on their surface. White can't help but wonder if there is any significance to the number of fossils Hillary used. Twelve on each side. Twelve disciples, perhaps? He must ask him. But they remind him of something else.

White stands looking at them for a few minutes, trying to remember where he had seen something similar. Then it comes to him. In the nearby village of Ashmansworth the old medieval church has on its inside walls engraved patterns consisting of circles, each about the size of a dinner plate. Within each circle are inscribed four petals that form a cross. Certainly, there are four rather than five petals, but otherwise they are very similar. He is pleased he has thought of this, especially today, for the pattern is known as the consecration cross; twelve were placed at intervals around the inside of the church, and probably another twelve on the outside. When the Ashmansworth church had been consecrated in the twelfth century, these crosses marked the points at which the walls of the church were anointed by the bishop during the ceremony. That was it! If the bishop asked, White could say the fossils were the locals' idea of consecration crosses. Time to go and look at what Hillary did to the little school. One of his workmen had told White at lunch that Mr. Hillary had outdone himself and put forty-six fossil urchins around the school's window. Much the same number of people that managed to fit inside the church for the consecration. One fossil for each of the villagers, perhaps? He'd have to find out.

Being so engrossed in Hillary's handiwork on the church, White doesn't hear the footsteps behind him. Turning, he comes face to face with the one person he would really rather like to have avoided—the bishop.

"Ah, Mr. White, I have been meaning to catch up with you all afternoon. You, sir, have captured the essence of the old church with your design."

Oh, dear, does he mean what I think he might? White thinks to himself, before thanking the bishop.

The bishop, much to White's relief, turns to leave. He suddenly stops. "Oh, yes, Mr. White. The ornament around the window."

White's intestines began to liquefy. *Dammit, he's noticed.*

"A nice touch. I'm sure our Lord, being a shepherd himself, rather likes the idea of a window crowned with shepherds' crowns."

ACKNOWLEDGMENTS

Over the more years than I care to remember that I have spent writing this book, it seemed to develop a life all its own, and it seemed to me that rather than I leading the book, it was often leading me, though usually down some very interesting pathways, fortunately. However, neither the book nor I could have reached the end without the help of many people. Throughout the writing process there was a small core of colleagues who at various times helped enormously, particularly in feeding me with relevant literature. I wish especially to thank David Reese for his encouragement and for loaning me specimens from Jordan to study, along with a constant supply of literature relevant to the Near East. Christian Neumann helped tremendously with relevant German literature. John Cooper of the Booth Museum of Natural History in Brighton was most helpful, and kindly allowed me the use of Herbert Toms's unpublished notes as well as an unpublished article on Toms and Pitt-Rivers. Chris Duffin kindly provided me with copies of his work on the mythology of sea urchin spines, and copies of Toms's photos of echinoids in windowsills. I am very grateful to Genevra Kornbluth for enlightening me on Merovingian bound pendants made with fossil urchins. Fiona Marsden of the Sussex Archaeological Society provided a number of useful references on fossil urchins in archaeological sites in Sussex. Didier Néraudeau supplied me with a preprint of his paper on French fossil urchins in archaeological sites, for which I am most grateful. Others who helped with literature include Francois Poplin, Jean Roman, Luc and Karinne Ceulemans, Jean Yves and Michelle Poirier, Renee Sipp, and Serge Rivault. The late Pat Vinnicombe shared her insight into Kenneth Oakley with me and provided me with relevant articles on the early evolution of art. In this area I am also indebted to Ellen Dissenayake for reprints of her articles and the loan of slides, including the Swanscombe Acheulian hand axe. James Dyer kindly sent me articles about Worthington Smith, and his excellent biography of Smith formed the basis for the chapter on him. Sylvia Hallam has been very helpful over the years, providing me with many useful archaeological references. Adrienne Mayor has given support, encouragement, and very helpful comments on one of the later versions of the manuscript.

I am grateful to the various curators who granted me access to specimens in their care: Ian Beavis at the Tunbridge Wells Museum; Angela Care Evans at the British Museum, David Lewis at the Natural History Museum, London; Chris Chippendale and John Osborne at the Archaeology and Anthropology Museum at the University of Cambridge; Gary Brown at the Liverpool Museum; and Robin Holgate, when at the Luton Museum, Bedfordshire, introduced me to Maud and provided me with literature and a photograph of Worthington Smith. Others who have given helpful advice are Gale Sievking, Mark Tosdevin, and David Tomalin.

I wish to thank a number of people who have provided me with photographs of specimens from either their own collections or collections in their care. First and foremost thanks to Roland Meuris, who with the help of John Jagt of the Natuurhistorisch Museum Maastricht, The Netherlands, sent me many superb slides of the Neolithic flint artifacts he had found that contain echinoids. Thanks also to Elisabeth Munksgaard of the Nationalmuseet, Copenhagen; Jan Peder Lamm at the Statens historiska museum, Stockholm; Didier Néraudeau at the Université de Rennes; Judith Bunbury at the University of Cambridge; and Enrichetta Leospo at the Museo Egizio, Turin.

Two people whom I wish especially to thank are Betty and Alan Smith at Linkenholt. They shared with me their knowledge of fossil urchins; provided me with invaluable information on the Linkenholt church, St. Peter's; and gave me a shepherd's crown and a shepherd's heart, both of which are illustrated in this book.

I am grateful to Simon Conway Morris for finding out that the star on the St. John's College Chapel organ is a cymbelstern, and to Simon Lawford for explaining just what it does. Thanks also to Simon Lawford for his help in producing such a great sound from the St. George's Cathedral Choir in Perth, Australia. Quite a lot of the writing of this book was carried out while listening to them as they rehearsed.

For help with translating I would like to thank Massimo Buonaiouto,

Marcos Serrano, Sten Vikner, and Malte Ebach. Malte and Caitlin Hulcup also very kindly read and commented on some early drafts of the chapters. At the University of Chicago Press I have been fortunate in having such an enthusiastic editor in Christie Henry. Without her support, this book would never have seen the light of day. I am most grateful to Sandra Hazel for doing such an excellent job in editing the final manuscript. Thanks also to Margaret Triffitt, librarian at the Western Australian Museum, for help over the years in tracking down many obscure articles for me.

And last, but most important, I must dedicate this book to my family: to John and Janet Groom, and Jean Radford, for their support. And to my wife, Sue, and children, Jamie, Katie, and Tim, whom I cannot thank enough. All have helped enormously with this book in many ways, and have had to live with it almost as much as I have over the last few years. Sue and Jamie in particular I wish to thank for reading and commenting on countless drafts.

NOTES

INTRODUCTION
1. Wynn 2002, p. 389.
2. Linnaeus 1758, vol. 1, p. 20.
3. Linnaeus 1806 (English ed.).
4. Linnaeus 1806, p. 9.
5. Muensterberger 1994.
6. Field 1965.

CHAPTER 1
1. Oakley 1985, p. 28.
2. Mithen 1994, p. 36.
3. Marean et al. 2007.
4. Henshilwood et al. 2002.
5. Bullington and Leigh 2002.
6. Mousterian is a tool technology mainly associated with Neanderthals between about two hundred thousand and forty thousand years ago.
7. Vértes 1964, pp. 141, 142.
8. Feliks 1998.
9. Ficatier 1886.
10. Oakley 1978, p. 237.
11. Dissanayake 1998.
12. Oakley 1965, p. 10.

CHAPTER 2
1. The specimen is housed in the collections of the Department of Archaeology & Anthropology, University of Cambridge.
2. Oakley 1981, p. 208; Feliks 1998.

3. Feliks 1998.
4. Bahn 1998, p. 86.
5. Eiseley 1964, pp. 270–71.
6. Lévi-Strauss 1969.
7. Oakley 1971, pp. 582–83; 1981, p. 209.
8. Oakley 1971, p. 582; see also Oakley 1985, fig. 25.
9. Poplin 1988, fig. B2.
10. Odin and Néraudeau 2001.
11. Néraudeau 2004.
12. Marshack 1989.
13. Oakley 1965, p. 10, fig. 1C.
14. Baudouin 1935.
15. Demnard and Néraudeau 2001.
16. Oakley 1985, p. 23.
17. White 1993, p. 294.
18. Belemnites are the internal "guard" from an extinct group of cephalopods, similar to the modern-day cuttlefish. The "guard" is a bullet-shaped structure made of calcite which helped stabilize the animal as it swam through the seas.
19. Boriskovskii 1956, pp. 113–14.
20. White 1992, p. 41.
21. Oakley 1978, p. 225.
22. Altena 1962.
23. Fraas 1878; Oakley 1978.
24. There is some uncertainty as to whether Woodward was born in 1665 or 1668.
25. Woodward 1728, pp. 11–12.
26. Duffin 2008, p. 30.
27. Goodyer 1655 in Gunther 1968, p. 655.
28. Duffin 2006; 2008.
29. Bruel 1632, p. 334.
30. Duffin 2006, p. 267; 2008, p. 31.
31. Wirtzung 1617, p. 456.
32. Lev and Amar 2000; 2002
33. Duffin 2008, p. 32.
34. England and Wales 1650, p. 30.
35. Duffin 2008, p. 33.
36. Maimonides 1211 in Muntner 1966, p. 14.
37. Kuhn et al. 2001, p. 7641.
38. Ibid., p. 7645.
39. Larsson 1983, p. 26.
40. Glørstad et al. 2004, pp. 101–2.

CHAPTER 3

1. Dean 1999.
2. Mantell 1822.
3. Thanks to Professor Hugh Torrens for sharing with me his interpretation of the distinction between those who actually find and pick up fossils (the hunters) and those

who acquire them for their collection through other means, such as purchase or barter (these are the gatherers).
4. McGowan 2001, p. 55.
5. Rudwick 2008.
6. Adams 1954, p. 12.
7. Mayor 2000.
8. Edwards 1976, p. 2.
9. Ibid.
10. Rudwick 1976, chap. 1.
11. Wendt 1968, p. 28.
12. Ibid.
13. Edwards 1976, p. 19.
14. Ibid., p. 21.
15. Rudwick 1976, pp. 56–58.
16. Zammit Maempel 1982, p. 22.
17. Scilla 1747, explanation to plate 24.
18. Woodward 1729, p. 8.
19. Friebe 1995; Thenius and Vávra 1996.
20. Adams 1954, p. 258.
21. Casaubon 1672, p. 124.
22. Drake 1996.
23. Rudwick 1976, p. 54.
24. Hooke 1665, p. 111.
25. Jardine 2003, chap. 2.
26. Waller 1705, plate 3; Drake 1996, p. 167.
27. Waller 1705, p. 284; Drake 1996, p. 166.

CHAPTER 4

1. Dean 1999, p. 73, n. 20.
2. Flower 1986, p. 77.
3. Curwen 1954, p. 162.
4. Violet Alford, quoted in Godman 1956, p. 171.
5. Curwen 1954.
6. Curwen 1934.
7. Wilson 1975, p. 183.
8. Whittle 1996, p. 274.
9. Obituary of Dr. Eliot Cecil Curwen, anon. in *Sussex Notes and Queries* 16, pp. 357–58.
10. Curwen 1934, p. 100.
11. Ibid., p. 103.
12. Ibid., p. 100.
13. Ibid., p. 101.
14. Curwen 1954, p. 83.
15. Curwen 1934, p. 102.
16. Ibid., p. 108.
17. Ibid., pp. 108, 110.

18. Bristow 1998, p. 30.
19. Curwen 1934, p. 110.
20. Curwen 1954, p. 81.
21. Ibid., pp. 81, 82.
22. Ibid., p. 82.
23. Ibid., p. 84.
24. Whittle et al. 1999.
25. Castleden 1987, p. 235.
26. Whittle 1996, p. 243.
27. Parker Pearson 1999, p. 158.
28. Miles Russell's foreword in Brook 2003.
29. Pull 2003a, p. 31. *Beaker* is the term used for a widely distributed culture in western Europe, extending from the late Neolithic to the early Bronze Age, about 2600 BC to 1900 BC. It is defined by the common use of a particular style of pottery that has a characteristic bell shape. Many archaeologists consider that Beaker people, per se, did not form a discrete ethnic group, but that the beakers and associated material are more indicative of the development of particular manufacturing skills.
30. White 2001, pp. 8, 9.
31. Pull 2003a, p. 32.
32. White 2001, p. 6.

CHAPTER 5

1. Brown 2007.
2. Wyse and Connolly 2002, pp. 139–43.
3. Oakley 1978, p. 223.
4. Rahtz 1971.
5. Rahtz 2002, p. 13.
6. Ibid., p. 16.
7. Donovan 1968.
8. Rahtz 2002, p. 16.
9. Whittle 1996, p. 1.
10. Ankel 1958, pp. 130, 131.
11. Gronenborn 1999.
12. Struve 1955.
13. Raymond 1907.
14. Morel 1971, 1977.
15. Ankel 1958, p. 133.
16. De Vesly 1903.
17. Clouet 1897, p. 95.
18. Du Chatellier 1907.
19. Chauvet 1900.
20. Holgate 1995.
21. Oakley 1985, plate 4c.
22. Ibid., p. 34.
23. Weiner et al. 1953.
24. Oakley 1965, p. 10.
25. Cayeux 1958.

26. Bellucci 1900, pp. 192–96.
27. As of September 2003, Roland Meuris had collected the following fossil urchins in artifacts: 64 *Cardiaster granulosus*, 22 *Micraster ciplyensis*, 5 *Echinocorys* sp. (Roland Meuris, personal communication, 2003).
28. Dyer 1978, p. 157.

CHAPTER 6

1. Dyer 1978, p. 154.
2. Ibid., p. 160.
3. Ibid., p. 144.
4. Ibid., p. 146.
5. Ibid., p. 142.
6. Large 1940, p. 167.
7. Dyer 1978, p. 144.
8. Smith 1894, p. 190.
9. White 1997, p. 312.
10. Ibid.
11. Smith 1894, pp. 273–74.
12. Bednarik 2005.
13. Oakley 1978, p. 227.
14. Dyer 1978, p. 155.
15. Ibid.
16. Smith 1894, p. 338.
17. Dyer 1978, p. 156.
18. Ibid., p. 155.
19. Smith 1894, p. 337.
20. Dyer 1978, p. 156.
21. Smith, quoted in ibid., p. 156.
22. Harding 2000, pp. 111–14.
23. Schlosser 2005.
24. Blunt 1950, p. 242.

CHAPTER 7

1. *Quoit* derives from the Middle English *coyte*, meaning a flat stone. Quoits are megalithic structures that consist of large standing stones supporting a flat capstone. They were probably covered by soil and were essentially a stone-supported barrow. One possibility for the name quoitstones is that because of the similarity of the shape of *Clypeus plotii* fossils to the capstone of the quoits, they were known as quoitstones. This would provide a further link between fossil urchins and barrows. Moreover, the game of "quoits" was sometimes played using a flat stone, and it is possible that Smith, as a boy, used these fossils in such a game.
2. Phillips 1844, p. 3.
3. Fitton 1833, p. 38.
4. Plot 1677, p. 91.
5. Darwin 1859.
6. Rowe 1899; Gale and Cleevely 1989.
7. Bowden 1991.

8. Ibid., p. 15.
9. Tylor 1922, p. 1141.
10. Pitt-Rivers 1891, p. 116.
11. Bowden 2004.
12. Bowden 1991, pp. 60–64.
13. Ibid., p. 31.
14. Bahn 1996, p. 56.
15. Ibid.
16. Ibid.
17. Pitt-Rivers 1869.
18. Bowden 1991, p. 95.
19. Pitt-Rivers 1888.
20. Holleyman 1987, p. 12.
21. Bowden 1991, p. 106.
22. Hawkins 1981, p. 141.
23. Pitt-Rivers 1888.
24. Ibid.
25. Ibid., p. 94.
26. Ibid., p. 98.

CHAPTER 8

1. Toms 1940.
2. Holleyman 1987, p. 13.
3. Duffin 2009, p. 57.
4. Ibid., p. 28.
5. Toms 1940, pp. 6, 7.
6. Ibid., p. 6.
7. From Toms's card index dated July 28, 1929, quoted by Duffin (2009).
8. Toms 1940, p. 6.
9. Ibid., p. 7.
10. Raymond 1907, pp. 136, 137.
11. Toms 1940, p. 2.
12. Ibid., p. 7.
13. Ibid.
14. Objects are "blackleaded" by polishing with "blacklead," or graphite.
15. Toms 1926.
16. These he listed at the last talk he ever gave on this subject to the Brighton Natural History Society, in January 1940, the year he died. Fortunately he made notes of this talk, which were typed up. I am grateful to John Cooper of the Booth Museum of Natural History in Brighton—another man, like Toms, fascinated by the folklore of these little objects—for providing me with a copy of Toms's talk.
17. Toms 1926, p. 264.
18. A saveloy is a highly seasoned pork sausage.
19. Oakley 1985, p. 26, fig. 24.
20. Plot 1677, p. 62.
21. Toms 1940, p. 8.
22. Betty Smith, personal communication, 2003.

23. Mantell 1844, p. 344.
24. Toms 1940, p. 8.
25. Keightley 1968.
26. Ibid., p. 8.
27. Oakley 1985, p. 27.
28. Evans 1966.
29. Squire 1905, p. 137.
30. Oakley 1985, p. 27.
31. Evans 1966, p. 129.
32. Mantell 1844, p. 350.
33. Rod Long, personal communication, 2009.
34. "Eotenas ond ylfe ond orcneas / swylce gigantas": *Beowulf*, lines 112–13. Translates as: "etins and elves and orcs / such giants."
35. McInnes 2002, p. 260.
36. Toms 1940, p. 4.
37. Rod Long, personal communication, 2009.
38. Toms's unpublished notes, Brighton Museum.
39. Field 1965.
40. Adrienne Mayor, personal communication.

CHAPTER 9

1. Harding 1994.
2. Heaney 1999, line 3155.
3. Ibid., lines 3157–58.
4. Ibid., p. 99.
5. Barrow 1855.
6. Ibid., p. 164.
7. Heaney 1999, lines 3160–63.
8. Pérot 1917, p. 100.
9. De Vesley 1903.
10. It was number 79 in a series entitled *Studier fra Sprog-og Oldtidsforskning udgivne af det philologisk-historiske Samfund*.
11. Blinkenberg 1911, pp. 68–82.
12. In Sweden the bullet-shaped belemnite rostra are called *vätteljus*. In Swedish folklore *vättar* is a collective word for creatures that are part elf or fairy and part the soul of deceased people. Some are good-natured, while others are decidedly evil. The strong magical powers of the *vätteljus* were thought to protect against evil, particularly at Christmastime, when more nasty *vättar* were thought to be about. A more modern (and child-friendly) interpretation views *vättar* as little creatures distantly related to Santa Claus. In this modern version, belemnites are thought to have been candlelight used by these tiny creatures to light their underground dwellings. It is considered very important to please these little *vättar*, who often live under the house, by keeping the house tidy. Otherwise, they may get very upset and might even turn on the children, making them sick (Mikael Siversson, personal communication, 2005).
13. Blinkenberg 1911, p. 81.
14. Ibid., p. 82.
15. Ibid., pp. 96, 97.

16. Ibid., p. 76.
17. Ibid., p. 82.
18. Sørensen 1994, fig. 2.
19. Blinkenberg 1911, pp. 82, 83.
20. Sørensen 1994.
21. I am indebted to Professor Sten Vikner of the Institute of Language, Literature & Culture at the University of Aarhus, Denmark, for help with translation.
22. Sørensen 1994, pp. 49–51.
23. Ibid., p. 49.
24. Ibid., p. 51.
25. Ibid., p. 48.
26. Blinkenberg 1911, p. 84.
27. Ibid.
28. Ibid.
29. It was from Lovett that Pitt-Rivers had bought the shepherd's crown that Lovett had obtained in Sussex (see p. 119).
30. I have to admit that when I was young and growing up in Sussex, after a storm we would rush outside and look for the thunderbolt. What we were looking for, though, wasn't a fossil, but what I now know to be nodules of iron pyrite that often occur in the chalk downland.
31. Blinkenberg 1911, p. 13.
32. The name literally means "thunder."
33. The short handle of Mjöllnir was said to be due to the dwarf who forged it having been interrupted by a gnat that stung him on the eyelid. But this was no ordinary gnat. It was Loki, the creator of mischief among the gods.
34. Blinkenberg 1911, p. 60.
35. This rhyme is derived from the old *Swedish Rhyme Chronicle*, p. 428.
36. Hollander 1928, pp. 89–90.
37. BBC news report, November 26, 2002.
38. Magnusson 1976, p. 82.
39. Wagner and Macdowall 1891, pp. 305–6.
40. Blinkenberg 1911, p. 96.
41. Wagner and Macdowall 1891, p. 146.

CHAPTER 10

1. Rollefson et al. 1992.
2. Ibid.
3. Ibid.
4. Schmandt-Besserat 1998.
5. Ibid.
6. Grissom 2000.
7. McNamara 2004.
8. Ibid.
9. McNamara 2002, figs. 12.25–27.
10. McNamara 2004.
11. Schmandt-Besserat 1998.
12. Szumowski 1955.

13. Demnard and Néraudeau 2001, fig. 8.
14. Bienkowski 2002, p. 42.
15. McNamara 2002.
16. Ibid., fig. 12.11.
17. Ibid., fig. 12.12.
18. Reese 2002.
19. McNamara 2002, p. 446.
20. Reese 1995.
21. Reese 2000.
22. Ibid.
23. Zammit 1930.
24. Ibid., p. 29.
25. Ibid., pp. 122–23.
26. McNamara, in press.
27. Price 1988.
28. Oakley 1985, p. 34.
29. Petrie 1933, plate 17.
30. De Morgan 1896.
31. Specimen number UC5935, Petrie Museum of Egyptian Archaeology, University of London.

CHAPTER 11

1. Shaw and Nicholson 1995, p. 45.
2. Ibid., pp. 275–76.
3. Anonymous 1996.
4. Scamuzzi 1947.
5. Schumacher 1988.
6. Susanne Binder (Tübingen), personal communication, 2005.
7. McDermott 2001, p. 148.
8. Edwards 1985, p. 9.
9. Fletcher 2002, p. 36.
10. Brewer and Teeter, 1999.
11. Parker 1974, p. 53.
12. Ibid., fig. 5.
13. Edwards 1985, p. 13.
14. Spence 2000, pp. 320–24.
15. McDermott 2001, p. 70.
16. Petrie 1895, p. 88.
17. Ibid., p. 89.
18. McDermott 2001, p. 112.
19. This translation from the Pyramid Texts, like subsequent ones in this book, is from *The Ancient Egyptian Pyramid Texts*, translated by R. O. Faulkner (Oxford: Clarendon Press, 1969).
20. Jean-François Champollion, quoted in Adkins and Adkins 2000, p. 177.
21. Ibid., p. 41.
22. Parker 1974, p. 56.
23. Edwards 1985, pp. 10–11.

24. Herodotus, quoted in Rundle Clark 1949, p. 1.
25. Rundle Clark 1949, p. 3.
26. Ibid., p. 3.
27. Ibid., p. 3, note 16; p. 15.
28. Ibid., pp. 6, 7.

CHAPTER 12

1. Suetonius, p. 185.
2. Grant 1975, p. 134.
3. Suetonius, p. 191.
4. Grant 1975, p. 138.
5. Suetonius, p. 191.
6. Ibid., p.191.
7. Pliny, translated by Holland 1634, book 29, chap. 13. This rendering is preferred to later nineteenth-century and twentieth-century translations, because it lacks omissions found in those editions and is generally a more accurate translation.
8. Breton et al. 2005.
9. It should be noted, though, that the objects that these were referring to were blue glass beads, rather than fossil urchins.
10. Pliny, translated by Holland 1634, book 37, chap. 10.
11. Adams 1954, p. 118.
12. Plot 1677, p. 90.
13. Brookes 1763.
14. Plott 1677, p. 106.
15. Ibid., pp. 106, 107.
16. Siculus 8 BC, *Histories* 5.28.6.
17. Strabo 10 AD, *Geographies*, bk. 4, chap. 4, p. 249.
18. Oakley 1985, fig. 19.
19. Pérot 1917, p. 101.
20. Kornbluth in press; Stanek 1999; Quast 2000.
21. Nationalmuseet, Copenhagen, no. 14172.
22. Stanek 1999, p. 361.
23. Ibid.
24. Sehested 1878, plate 35r.
25. Oakley 1985, p. 32.
26. Yliali Asp, Senior Curator, Department of Documentation, Staatsmuseum Sweden, personal communication, 2001.
27. Metzger-Krahé 1978, p. 41.
28. Ibid., p. 46.
29. Guébhard 1907, p. 344.
30. Gilchrist 2008, p. 149.
31. Ibid., p. 133.
32. Douglas 1785.
33. Dean 2004.
34. Douglas 1793, p. 65, caption to plate 15, fig. 14.
35. Ibid.
36. West 1988, p. 32, figs. 42, 73H.

Notes to Pages 200–209

37. Buxton 1921.
38. Schmidt 1970, plate 3, fig. 2.
39. Meaney 1981, p. 117.
40. British Museum, specimen number 1963, 1108.334. From grave 55.

CHAPTER 13

1. Ellis Davidson 1964, p. 217.
2. Ibid.
3. *Wyrd* is pronounced "wurd."
4. Branston 1957, p. 65.
5. It is rather significant in this context, perhaps, that a living heart urchin occurs in the Indo-Pacific region, was described by Lamarck in 1816 and is called *Moira atropos*.
6. Branston 1957, p. 79.
7. Graves 1955, p. 48.
8. Ibid., p. 48.
9. Ibid., pp. 9–10.
10. Branston 1957, p. 79.
11. Graves 1955, p. 11.
12. Taylor and Auden 1969.
13. Branston 1957, p. 59.
14. Ellis Davidson 1964, p. 217.
15. Heaney 1999, p. xiv.
16. Ibid., p. xvii.
17. Ibid., lines 2569–70.
18. Ibid., line 2814.
19. Alexander 1966, p. 30.
20. Ibid., p. 67.
21. Ibid., p. 70.
22. Hamer 1970, p. 175.
23. Ibid.
24. Ibid., p. 181.
25. Branston 1957, p. 59.
26. Plato, *Republic*, 616b.
27. Ibid., 617b,c
28. Plato, *Timaeus*, 42B.
29. Swanton 1975, pp. 116–25.
30. Allen 1941, p. 164.
31. *King Lear*, act 4, scene 3, lines 34–35.
32. Allen 1941, p. 153.
33. Ibid., p. 161.
34. Ibid.
35. Barber 1991, p. 52.
36. Demnard and Néraudeau 2001.
37. Gilchrist 2008, p. 133.
38. Burchard of Worms ca. 1010, quoted in ibid.
39. Richard Hobbs, British Museum, personal communication, 2009.
40. Oakley 1985, fig. 20.

41. Oakley 1978, p. 233.
42. McNamara in press.
43. Williams 2008, p. 56.
44. Akerman 1856, p. 258.
45. Wylie 1852.
46. Akerman 1855, plate 19, fig. 5.

CHAPTER 14

1. Stone 1974, p. 44–46.
2. Translation by the author.
3. Usually the five-pointed star symbol is referred to as a pentagram when used in a pagan context, or as a pentacle when inscribed within a circle, which makes it look for all the world like a fossil sea urchin. Pentagrams are usually represented not as solid stars, but by a single line.
4. BBC news report, November 26, 2002.
5. Biedermann 1993.
6. Hulme 1969, p. 213.
7. Webber 1992, pp. 26, 73, 382.
8. Lehner 1950, p. 582.
9. De Vogel 1966.
10. Mackey 2002.
11. Schouten 1968, p. 23, fig. 3.7.
12. Ibid., p. 21, fig. 3.6.
13. Ibid., p. 23.
14. Ibid., p. 24.
15. Ibid.
16. Ibid., pp. 17–18.
17. Ibid., p. 52, fig. 3.22.
18. The word *star* itself has an ancient heritage, suggesting that the coincidence of using this word for the pattern and the heavenly body occurred a very long time ago. The lineage is shown by the close similarity of the word for *star* among many Indo-European languages that are seemingly very different today, but which are thought to have had a common origin, between 9,800 to 11,800 years ago. Here are a few words for *star* in a number of these languages, in no particular order:

Modern English	star
Old English	sterre
Anglo-Saxon	steorra
Old Friesian	stera
Old Saxon	sterro
Dutch	ster
Old High German	sterno, sterro
German	stern
Icelandic & Old Norse	stjarna
Swedish	stjerna
Danish	stierne
Gothic	stairno
Armorican	steren

Cornish	*steren*
Welsh	*s(t)eren*
Latin	*stella*
Greek	*aster*
Spanish	*estrella*
Hindi	*sitharee*
Sanskrit	*stri*
Avestan (Persian)	*star*
Middle Persian	*star*
Modern Persian	*stare*
Egyptian	*saba*

19. Van Buren 1945; Goff 1963.
20. Matthews 2002.
21. Matthews 1997.
22. Ashmolean Museum, Oxford, specimen number AN 1954.738.
23. Schouten 1968.
24. De Vogel 1966, p. 45.
25. Schouten 1968, p. 33, fig. 3.14.
26. Ibid., p. 31.
27. Ibid., pp. 31, 33.
28. *East Anglican Daily Times*, March 1, 2002.
29. Fisher 1970, p. 158.
30. Tyack 1899.
31. Fisher 1970.
32. Information from a plaque at the church entrance.

EPILOGUE

1. *The Marlborough Times and Wilts and Berks County Paper*, Friday, October 6, 1871.

REFERENCES

Adams, F.D. 1954. *The Birth and Development of the Geological Sciences*. New York: Dover.
Adkins, L., and R. Adkins 2000. *The Keys of Egypt: The Race to Read the Hieroglyphs*. London: HarperCollins.
Akerman, J. Y. 1855. *Remains of Pagan Saxondom*. London: J. R. Smith.
———. 1856. Several objects discovered in the Anglo-Saxon Cemetery at Fairford, Gloucestershire. *Proceedings of the Society of Antiquaries of London* 4:258.
Alexander, M. 1966. *The Earliest English Poems*. London: Penguin Classics.
Allen, D. C. 1941. *The Star-Crossed Renaissance: The Quarrel about Astrology and Its Influence in England*. Durham, NC: Duke University Press.
Altena, C. O. van R. 1962. Molluscs and echinoderms from Palaeolithic deposits in the rock shelter of Ksar' Akil, Lebanon. *Zoologische Mededelingen Rikjsmus* 38:87–99.
Ankel, C. 1958. Ein fossiler Seeigel vom Euzenberg bie Duderstadt (Süd Hanover). *Die Kunde: Niedersächsischer Landesverein für Urgeschichte; Sonerdruck* 9:130–35.
Anonymous. 1996. Giant of Egyptology: Ernesto Schiaparelli. *KMT: A Modern Journal of Ancient Egypt* 7, no. 2:82–84.
Bahn, P. G. 1996. *The Cambridge Illustrated History of Archaeology*. Cambridge: Cambridge University Press.
———. 1998. *The Cambridge Illustrated History of Prehistoric Art*. Cambridge: Cambridge University Press.
Bandel, K., and J. F. Geys. 1985. Regular echinoids in the Upper Cretaceous of the Hashemite Kingdom of Jordan. *Annales de la Societé Géologique du Nord* 54:97–115.
Barber, E. J. W. 1991. *Prehistoric Textiles*. Princeton, NJ: Princeton University Press.
Barrow, B. 1855. Notes on the opening of the tumuli on Ashey Down. *Journal of the British Archaeological Association* 10:162–65.
Bassett, M. G. 1982. *Formed Stones—Folklore and Fossils*. Cardiff: National Museum of Wales.
Bates, B. 2002. *The Real Middle Earth: Magic and Mystery in the Dark Ages*. London: Sidgwick and Jackson.

Baudouin, M. 1935. Les oursins fossils travaillés: Grandes cupelettes et perforations. In *Congrés International d'Anthropologie et Archéologie Préhistorique, Bruxelles*, 220–39.

Bednarik, R. G. 2005. Middle Pleistocene beads and symbolism. *Anthropos* 100:537–52.

Bellucci, G. 1900. Echini mesozoici utilizzati dall'uomo dell'età della pietra. *Bullettino di Paletnologia Italiana* 26:193–96.

Biedermann, H. 1993. *Dictionary of Symbolism: Cultural Icons and the Meaning behind Them*. New York: Facts on File.

Bienkowski, P. 1995. The small finds. In *Excavations at Tawilan in Southern Jordan*, ed. C. M. Bennett and P. Bienkowski, 79–92. British Academy Monographs in Archaeology 8. Oxford: Oxford University Press.

———, ed. 2002. *Busayra Excavations by Crystal-M. Bennett 1971–1980*. British Academy Monographs in Archaeology 13. Oxford: Oxford University Press.

Black, J. 2002. Ancient Mesopotamia. In *Mythology: The Illustrated Anthology of World Myth and Storytelling*, ed. C. S. Littleton, 82–133. San Diego: Thunder Bay Press.

Blanckenhorn, M. 1925. Die Seeigelfauna der Kreide Palästinas. *Paläontographica* 67:83–113.

Blinkenberg, C. 1911. *The Thunderweapon in Religion and Folklore: A Study in Comparative Archaeology*. Cambridge: Cambridge University Press.

Blunt, W. 1950. *The Art of Botanical Illustration*. London: Collins.

Boriskovskii, P. I. 1956. Belemnity v drevnem kamennom veke'. *Priroda* 11:113–14.

Bowden, M. 1991. *Pitt Rivers: The Life and Archaeological Work of Lieutenant-General Augustus Henry Lane Fox Pitt Rivers, DCL, FRS, FSA*. Cambridge: Cambridge University Press.

———. 2004. Rivers, Augustus Henry Lane Fox Pitt- (1827–1900). In *Oxford Dictionary of National Biography*. Oxford: Oxford University Press.

Branston, B. 1957. *The Lost Gods of England*. London: Thames and Hudson.

Breton, G., F. Huet, and D. Vizcaino. 2005. Le coeur, l'étymologie et les oursins fosiles. *Bulletin de la Société géologique de Normandie et des amis du Muséum du Havre* 92:37–38.

Brewer, D. J., and E. Teeter. 1999. *Egypt and the Egyptians*. Cambridge: Cambridge University Press.

Bristow, P. H. W. 1998. Attitudes to disposal of the dead in southern Britain 3500BC–AD43. BAR British series, vol. 2, no. 274.

Brook, A., ed. 2003. John Henry Pull and Shepherds' Crowns. West Sussex Geological Society Occasional Publication no. 3.

Brookes, R. 1763. *A new and accurate system of natural history*. London: J. Newbery.

Brown, A. 2007. Dating the onset of cereal cultivation in Britain and Ireland: The evidence from charred cereal grains. *Antiquity* 81:1042–52.

Bruel, W. 1632. *Praxis medicinae, or, the physicians practice wherein are contained inward diseases from the head to the foote: explayning the nature of each disease, with the part affected; and also the signes, causes, and prognostiques, and likewise what temperature of the ayre is most requisite for the patients abode, with direction for the diet he ought to obserue, together with experimentall cures for euery disease. Practised and approved of: and now published by the good, not onely of Physicians, Chirurgions, and Apothecaries, but very meete and profitable for all such which are silicitous of their health and welfare*. London: John Norton.

Bullington, J., and S. R. Leigh. 2002. Rock art revisited. *Science* 296:468.

Buxton, L. H. D. 1921. Excavations at Frilford. *Antiquaries Journal* 1:87–97.

Casaubon, M., 1672. *A treatise proving spirits, witches, and supernatural operations, by pregnant instances and evidences together with other things worthy of note*. London: Brabazon Aylmer, at the Three Pigeons in Cornhill.

Castleden, R. 1987. *The Stonehenge People: An Exploration of Life in Neolithic Britain 4700–2000 BC*. London: Routledge and Kegan Paul.

Cayeux, L. 1958. Utilisation des fossils dans les industries préhistorique: Les oursins percuteurs de l'atelier des Sapinières (Fôret de Montgeon). *Bulletin Société Préhistorique Française* 55:135–36.

Chauvet, G. 1900. Ovum anguinum. *Revue Archéologique* 1:281–85.

Clouet, M. 1897. Découverte de la Garenne, commune de Juicq. *Recueil de la Commission des Arts et Monuments Historiques de la Charente Inférieure* 4th ser., 14:95–96.

Conway, B., J. McNabb, and N. Ashston, eds. 1996. *Excavations at Barnfield Pit, Swanscombe, 1968–1972*. British Museum, Occasional Papers 94.

Cotteau, G. 1869. Notice sur les echinides fossiles recueilles par M. Louis Lartet en Syrie et en Idumée, pendant son voyage avec Duc le Luynes. *Bulletin Société Géologique de France*, 2nd ser., 26:533–38.

Curwen, E. C. 1934. Excavations in Whitehawk Neolithic Camp, Brighton. *Antiquities Journal* 14:99–133.

———. 1954. *The Archaeology of Sussex*. Methuen and Co., London.

Darwin, C. 1859. *On the Origin of Species by Means of Natural Selection, or the Preservation of Favoured Races in the Struggle for Life*. London: John Murray.

Dean, D. R. 1999. *Gideon Mantell and the Discovery of Dinosaurs*. Cambridge: Cambridge University Press.

———. 2004. Douglas, James (1753–1819). In *Oxford Dictionary of National Biography*, online ed., http://www.oxforddnb.com/view/article/7902 (accessed April 26, 2009).

Demnard, F., and D. Néraudeau. 2001. L'utilisation des oursins fossiles de la Préhistoire à l'époque gallo-romaine. *Bulletin de la Société préhistorique Française* 98: 693–715.

Dissanayake, E. 1998. The beginnings of artful form. *Surface Design Journal* (Winter 1998): 4–6.

———. 2001. Birth of the arts. *Natural History* (December 2000–January 2001): 84–91.

Donovan, D. T. 1968. The ammonites and other fossils from Aveline's Hole (Burrington Combe, Somerset). *Proceedings of the University of Bristol Spelaeological Society* 11:237–42.

Douglas, J. 1785. *Dissertation on the Antiquity of the Earth, Read at the Royal Society, 12th May, 1785*. London: George Nicol.

———. 1793. *Nenia Britannica; or, A Sepulchral History of Great Britain . . .* London: Benjamin and John White.

Drake, E. T. 1996. *Restless Genius: Robert Hooke and His Earthly Thoughts*. Oxford: Oxford University Press.

du Chatellier, P. 1907. *Les Époques Préhistoriques et Gauloises dans la Finistère*. Paris: Rennes et Quimper.

Duffin, C. J. 2006. Lapis Judaicus or the Jews' stone: The folklore of fossil echinoid spines. *Proceedings of the Geologists' Association* 117:265–75.

———. 2008. Fossils as drugs: Pharmaceutical palaeontology. *Ferrantia* 54. Luxembourg: Musée national d'histoire naturelle.
———. 2009. Herbert Toms and the geological folklore of Sussex. *Journal of West Sussex History* 77:57–64.
Dyer, J. 1978. Worthington George Smith. *Publications of the Bedfordshire Historical Record Society* 57:141–77.
Edwards, I. E. S. 1985. *The Pyramids of Egypt*. 2nd rev. ed. London: Penguin.
Edwards, W. N. 1976. *The Early History of Palaeontology*. London: British Museum (Natural History).
Eiseley, L. 1978. *The Star Thrower*. New York: Harcourt.
Ellis Davidson, H. R. 1964. *Gods and Myths of Northern Europe*. London: Penguin.
England and Wales. 1650. *An Act for the Redemption of Captives*. London: Husband and Field (printers), Parliament of England.
Evans, G. E. 1966. *The Pattern under the Plough: Aspects of the Folk-life of East Anglia*. London: Faber and Faber.
Evans, J. 1897. *The Ancient Stone Implements, Weapons and Ornaments of Great Britain*. London: Longmans, Green and Co.
Faulkner, R. O. 1969. *The Ancient Egyptian Pyramid Texts*. Oxford: Clarendon Press.
Feliks, J. 1998. The impact of fossils on the development of visual representation. *Rock Art Research* 15:109–34.
Ficatier, A. 1886. Etude Paléolithiques sur la Grotte Magdalenienne du Trilobite à Arcy-sur-Cure (Yonne). In *Almanach historique du l'Yonne*, ed. A. Gallet. Auxerre: A. Gallet.
Field, N. H. 1965. Fossil sea-echinoids from a Romano-British site. *Antiquity* 39:298.
Fisher, E. A. 1970. *The Saxon Churches of Sussex*. Newton Abbot: David and Charles.
Fitton, W. H. 1833. Notes on the history of English geology. *London and Edinburgh Philosophical Magazine and Journal of Science* 2:37–56.
Fletcher, J. 2002. Egypt's divine Kingship. In *Mythology: The Illustrated Anthology of World Myth and Storytelling*, ed. C. S. Littleton, 10–81. London: Duncan Baird.
Flower, R. 1986. *The Old Ship: A Prospect of Brighton*. London: Croom Helm.
Fraas, O. 1878. Geologisches aus dem Libanon. *Jahrbuch des Vereins für Vaterländische Naturkunde in Württemburg* 34:257–81.
Frazer Hearne, E. J. 1934. Shepherds Garden, Arundel Park, a pre-Roman and Roman British settlement. *Sussex Archaeological Collection* 75:214–75.
Friebe, J. G. 1995. *Schlangeneier und Drachenzungen: Fossilien in Volksmedizin und Abwehrzauber*. Beiheft zur Sonderausstellung (September 23, 1995–January 1996): 44. Dornbirn: Vorarlberger Naturschau.
Gale, A. S., and R. J. Cleevely. 1989. Arthur Rowe and the Zones of the White Chalk of the English coast. *Proceedings of the Geologists' Association* 100:419–31.
Gilchrist, R. 2008. Magic for the dead? The archaeology of magic in later medieval burials. *Medieval Archaeology* 52:119–59.
Glørstad, H., H. A. Nakrem, and V. Tørhaug. 2004. Nature in society: Reflections over a Mesolithic sculpture of a fossilized shell. *Norwegian Archaeological Review* 37:95–110.
Godman, S. 1956. Good Friday skipping. *Folklore* 67:171–74.
Goff, B. L. 1963. *Symbols of Prehistoric Mesopotamia*. New Haven, CT: Yale University Press.
Grant, M. 1975. *The Twelve Caesars*. London: Weidenfield and Nicholson.
Graves, R. 1955. *The Greek Myths*. 2 vols. Harmondsworth, UK: Penguin.

Grissom, C. A. 2000. Neolithic statues from 'Ain Ghazal: Construction and form. *American Journal of Archaeology* 104:25–42.

Gronenborn, D. 1999. A variation on a basic theme: The transition to farming in southern central Europe. *Journal of World Prehistory* 13:123–210.

Guébhard, A. 1907. Sur l'universalité des superstitions attachées aux coquilles fossiles. *Bulletin de la Société Préhistorique Française* 4:344–45.

Gunther, R. T. 1968 [1933]. *The Greek Herbal of Dioscorides illustrated by a Byzantine A.D. 512, Englished by John Goodyer A.D. 1655.* London: Hafner.

Hamer, R. F. S. 1970. *Choice of Anglo-Saxon Verse: Parallel Text.* London: Faber and Faber.

Harding, A. F. 1994. Reformation in barbarian Europe. In *The Oxford Illustrated Prehistory of Europe*, ed. B. Cunliffe, 304–35. Oxford: Oxford University Press.

———. 2000. *European Societies in the Bronze Age.* Cambridge: Cambridge University Press.

Hawkins, D. 1981. *Cranborne Chase.* London: Victor Golancz.

Heaney, S., trans. 1999. *Beowulf.* London: Faber and Faber.

Henshilwood, C. S., F. d'Errico, R. Yates, Z. Jacobs, C. Tribolo, G. A. T. Duller, N. Mercier, et al. 2002. Emergence of modern human behavior: Middle Stone Age engravings from South Africa. *Science* 295:1278–80.

Holgate, R. 1995. Neolithic flint mining in Britain. *Archaeologia Polona* 33:133–61.

Hollander, L. M. 1928. *The Poetic Edda: Translated with an Introduction and Explanatory Notes.* Austin: University of Texas Press.

Holleyman, G. A. 1987. *Two Dorset archaeologists in Sussex: Lieut. General Pitt-Rivers in Sussex 1867–1878 and Herbert Samuel Toms Curator of the Brighton Museum 1896–1939.* Henley, West Sussex: Privately printed.

Hooke, R. 1665. *Micrographia; or, Some Physiological Descriptions of Minute Bodies Made by Magnifying Glasses, with Observations and Inquiries Thereupon.* London: Jo. Martyn and Ja. Allestry, printers to the Royal Society, the Bell in S. Paul's Church-yard.

Hulme, F. E. 1969. *The History of Principles and Practices of Symbolism in Christian Art.* Detroit: Gale Research Company.

Jardine, L. 2003. *The Curious Life of Robert Hooke: The Man Who Measured London.* London: HarperCollins.

Jessup, R. 1970. *Ancient People and Places: South East England.* London: Thames and Hudson.

Keightley, T. 1968. *The Fairy Mythology: Illustrative of the Romance and Superstition of Various Countries.* New York: Haskell House.

Kornbluth, G. In press. *Amulets, Power, and Identity in Early Medieval Europe.* Oxford: Oxford University Press.

Kuhn, S. L., M. C. Stiner, D. Reeses, and E. Güleç. 2001. Ornaments of the earliest Upper Paleolithic: New insights from the Levant. *Proceedings of the National Academy of Sciences* 98:7641–46.

Large, E. C. 1940. *The Advance of the Fungi.* London: J. Cape.

Larsson, L. 1983. Skateholmsprosjektet. Jägare—fiskare—bonder. *Limhamnia* 1983:7–40.

Lehner, E. 1950. *Symbols, Signs and Signets.* New York: World Publishing.

Lev, E., and Z. Amar. 2000. Ethnopharmacological survey of traditional drugs sold in Israel at the end of the 20th century. *Ethnopharmacology* 72:191–205.

———. 2002. Ethnopharmacological survey of traditional drugs sold in the Kingdom of Jordan at the end of the 20th century. *Ethnopharmacology* 82:131–45.

Lévi-Strauss, C. 1969. *The Raw and the Cooked*. In *Mythologiques*, vol. 1. Chicago: University of Chicago Press.

Levine, J. M. 2004. Woodward, John (1665/1668–1728). In *Oxford Dictionary of National Biography*, online ed., http://www.oxforddnb.com/view/article/29946 (accessed April 24, 2009).

Linnaei, C. 1758. *Systema naturae per Regna Tria Naturae, secundum classes, ordines, genera, species, cum characteribus, differentiis, synonymis, locis.* 2 vols. Holmiae: Laurenti Salvii.

———. 1806. *A general system of naturae, through the three grand kingdoms of animals, vegetables, and minerals, systematically divided into their several classes, orders, genera, species and varieties, with their habitations, manners, economy, structure, and peculiarities.* Vol. 1, *Mammalia, Birds, Amphibia, Fishes*, translated by William Turton. London: Lockington, Allen, and Co.

Mackey, A. 2002. *Encyclopedia of Freemasonary*. Whitefish, MT: Kessinger Publishing.

Magnusson, M. 1976. *Hammer of the North: Myths and Heroes of the Viking Age*. London: Orbis.

Mantell, G. 1822. *Fossils of the South Downs; or, Illustrations of the Geology of Sussex*. London: L. Relfe.

———. 1844. *The Medals of Creation*. London: I. Bohn.

Marean, C. W., M. Bar-Matthews, J. Bernatchez, E. Fisher, P. Goldberg, A. I. R. Herries, Z. Jacobs, et al. 2007. Early human use of marine resources and pigment in South Africa during the Middle Pleistocene. *Nature* 449:905–8.

Marshack, A. 1989. Evolution of the human capacity: The symbolic evidence. *Yearbook of Physical Anthropology* 32:1–34.

Matthews, R. J. 1997. *The Oxford Encyclopedia of Archaeology in the Near East*. American Schools of Oriental Research, vol. 3. Oxford: Oxford University Press.

———. 2002. *Secrets of the Dark Mound: Jemdet Nasr 1926–1928*. Warminster: British School of Archaeology in Iraq.

Mayor, A. 2000. *The First Fossil Hunters: Paleontology in Greek and Roman Times*. Princeton, NJ: Princeton University Press.

McDermott, B. 2001. *Decoding Egyptian Hieroglyphs*. London: Duncan Baird.

McGovern, P. 1983. Test soundings of archaeological and resistivity survey results at Rujm Al-Henu. *Annual of the Department of Antiquities of Jordan* 27:105–41.

McGowan, C. 2001. *The Dragon Seekers*. Reading, MA: Perseus.

McInnes, J. 2002. Celtic deities and heroes. In *Mythology: The Illustrated Anthology of World Myth and Storytelling*, ed. C. S. Littleton, 248–73. San Diego: Thunder Bay Press.

McNamara, K. J. 2002. Fossil marine invertebrates—echinoids. In *Busayra Excavations by Crystal-M. Bennett*, ed. P. Bienkowski, 442–54. British Academy Monographs in Archaeology 13. Oxford: Oxford University Press.

———. 2004. Fossil echinoids from Neolithic and Iron Age sites in Jordan. In *Echinoids: Munich*, ed. T. Heinzeller and J. Nebelsick, 459–46. Rotterdam: Balkema.

———. 2007. Shepherds' crowns, fairy loaves and thunderstones: The mythology of fossil echinoids in England. In *Myth and Geology*, ed. L. Piccardi and W. B. Masse, 279–94. Special Publication, Geological Society of London.

———. In press. Fossil sea urchins from Wadi ath-Thamad Site 13. In *Excavations at Wadi ath-Thamad, Jordan*, vol. 2, *The Iron Age Artefacts*. Leiden: Brill.

Meaney, A. L. 1981. Anglo-Saxon amulets and curing stones. BAR British ser. 96.

Metzger-Krahé, F., in B. Arrhenius et al. 1978. *Das archäologische Fundmaterial III der Ausgrabung Haithabu*. Schleswig-Holsteinisches Landesmuseum für Vor- und Frühgeschichte, Schleswig, Schloss Gottorp. Berichte über die Ausgrabungen in Haithabu. Bericht 12.

Mithen, S. 1994. From domain specific to generalized intelligence: A cognitive interpretation of the Middle/Upper Palaeolithic transition. In *The Ancient Mind: Elements of Cognitive Archaeology*, ed. C. Renfrew and E. B. W. Zubrow, 29–39. Cambridge: Cambridge University Press.

Morel, J. 1971. Oursins fossils perforés de la Saintonge. *Bulletin de la Société Préhistorique Française* 68:281–88.

———. 1977. Encore les oursins fossils perforés. *Bulletin de la Société Préhistorique Française* 74:213–16.

Morgan, J. de. 1896. *Recherches sur les origines d'Egypte: L'âge de la Pierre et les métaux*. Paris: E. Leroux.

Muensterberger, W. 1994. *Collecting, an Unruly Passion: Psychological Perspectives*. San Diego: Harcourt Brace.

Mutner, S., trans. 1966. *The Medicinal Writings of Moses Maimonides*. Vol. 2, *Treatise on Poisons and Their Antidotes*. Philadelphia: J. B. Lippincott.

Néraudeau, D. 2004. Les silex fossilifères du nord du littoral Charentais et leur utilisation au Paléolithique. *Bulletin A.M.A.R.A.I.* no. 17.

Neumann, C. 1999. Irregular echinoids from the Ajlun Group (Upper Cretaceous) of Jordan. In *Echinoderm Research 1998*, ed. M. D. Candia Carnevali and F. Bonasoro, 361–66. Rotterdam: Balkema.

Oakley, K. P. 1965. Folklore of fossils. *Antiquity* 39:9–16, 117–25.

———. 1971. Fossils collected by the earlier Palaeolithic men. *Mélanges de Préhistoire, d'archeocivilisation et d'ethnologie offerts à André Varagnac* 581–84. Paris: Sevpen.

———. 1978. Animal fossils as charms. In *Animals in Folklore*, ed. J. R. Porter and W. M. S. Russell, 208–40. Ipswich, UK: D. S. Brewer Ltd., Rowman and Littlefield for The Folklore Society.

———. 1981. Emergence of higher thought 3.0–0.2 Ma B.P. *Philosophical Transactions of the Royal Society of London* B 292:205–11.

———. 1985. *Decorative and Symbolic Uses of Fossils*. Occasional Papers on Technology 13. Oxford: Pitt Rivers Museum, University of Oxford.

Odin, G. S., and Néraudeau, D. 2001. Sur un nucléus préhistorique comportant un oursin fossile à Tercis. In *The Campanian-Maastrichtian Boundary*, ed. G. S. Odin, 23–26. Amsterdam: Elsevier.

Owen, E. 1987. Introduction. In *Fossils of the Chalk*, ed. A. B. Smith, 9–14. London: Palaeontological Association.

Parker, R. A. 1974. Ancient Egyptian astronomy. *Philosophical Transactions of the Royal Society of London* A 276:51–65.

Parker Pearson, M. 1999. *The Archaeology of Death and Burial*. Stroud, UK: Sutton.

Pérot, F. 1917. La survivance de l'oursin fossile. *Bulletin de la Societé Préhistorique Française* 14:100–102.

Petrie, W. M. F. 1895. *Egyptian Decorative Art: A Course of Lectures Delivered at the Royal Institution*. London: Methuen.

———. 1933. *Ancient Gaza III: Tell el Ajjul*. London: British School of Archaeology in Egypt.

Phillips, J., ed. 1844. *Memoirs of William Smith*. London: J. Murray.
Pitt-Rivers, A. Lane-Fox. 1869. An examination into the character and probable origin of the hill forts of Sussex. *Archaeologia* 42:2–52.
———. 1873–75. On the evolution of culture. *Notices of the Proceedings at the Meetings of the Members of the Royal Institution* 7:496–520.
———. 1888. *Excavations in Cranborne Chase, near Rushmore, on the borders of Dorset and Wilts.* Vol. 2, *Excavations in Barrows, near Rushmore; in Romano-British village, Rotherley; in Winkelbury Camp; in British Barrows and Anglo-Saxon Cemetery, Winkelbury Hill*. London: Privately printed (Harrison and Sons).
———. 1891. Typological museums. *Journal of the Society of Arts* 40:115–22.
Pliny the Elder. 1634. *The historie of the world: Commonly called, The naturall historie of C. Plinius Secundus.* Translated into English by Philemon Holland Doctor of Physicke. The first [-second] tome. London: Adam Islip.
Plot, R. 1677. *The Natural History of Oxford-shire, being an Essay toward the Natural History of England.* Oxford & London: Printed at the Theater in Oxford, and are to be had there: and in London at Mr. Moses Pits at the Angel in St. Pauls Church-yard, and at Mr. S. Millers, at the Star near the west-end of St. Pauls Church.
Poplin, F. 1988. Aux origines néandertaliennes de l'Art: Matière, forme, symétries; Contribution d'une galène et d'un oursin fossile taillé de Merry-sur-Yonne (France). *L'Homme de Néandertal, Liège*, 5:109–16.
Price, D. 1988. Minerals and fossils. In *The Egyptian Mining Temple at Timna*, ed. B. Rothenberg, 266–67. London: Institute for Archaeo-Metallurgical Studies.
Pull, J. H. 2003a. Shepherds' Crowns—their occurrence as symbols in prehistoric graves. West Sussex Geological Society Occasional Publication no. 3, pp. 31–32.
———. 2003b. Shepherds' Crowns—the survival of belief in their magical virtues in Sussex. West Sussex Geological Society Occasional Publication no. 3, pp. 33, 35.
Quast, D. 2000. Amulett? Heilmittel? Schmuck? Unauffällige Funde aus Oberflacht. *Archäologisches Korrespondenzblatt* 30:279–94.
Rahtz, P. 1971. Excavations on Glastonbury Tor. *Archaeology Journal* 127:1–81.
———. 2002. Glastonbury Tor: A modified landscape. *Landscapes* 3:4–18.
Raymond, P. 1907. L'oursin fossile et les idées religieuses à l'époque préhistorique. *La Revue Préhistorique, Annales de Paléoethnologie* 2:133–39.
Reese, D. 1995. The invertebrate fossils in the physical geology of the western Mesara and Kommos. In *Kommos I/1: The Kommos Region, Ecology, and Minoan Industries*, ed. J. W. Shaw and M. C. Shaw, 87–90. Princeton, NJ: Princeton University Press.
———. 2000. Fossils. In *Kommos IV: The Greek Sanctuary*, ed. J. W. Shaw and M. C. Shaw, 403–7. Princeton, NJ: Princeton University Press.
———. 2002. Shells and fossils from Tall Jawa, Jordan. In *Excavations at Tall Jawa, Jordan*, vol. 2, *The Iron Age Artefacts*, ed. P. M. M. Daviau, 276–91. Culture and History of the Ancient Near East 11, no. 2. Leiden: Brill.
Roe, D. A. 1981. *The Lower and Middle Palaeolithic Periods in Britain*. London: Routledge and Kegan Paul.
Rollefson, G. O., A. H. Simmons, and Z. Kafafi. 1992. Neolithic cultures at 'Ain Ghazal, Jordan. *Journal of Field Archaeology* 19:443–70.
Rowe, A. W. 1899. An analysis of the genus *Micraster*, as determined by rigid zonal collecting from the zones of *Rhynchonella cuvieri* and *Micraster coranguinum*. *Quarterly Journal of the Geological Society of London* 55:494–547.

Rudwick, M. J. S. 1976. *The Meaning of Fossils: Episodes in the History of Palaeontology*. New York: Science History Publications.

———. 2008. *Worlds before Adam: The Reconstruction of Geohistory in the Age of Reform*. Chicago: University of Chicago Press.

Rundle Clark, R. T. 1949. The origin of the Phoenix: A study in Egyptian religious symbolism. Part 1, The Old Empire. *University of Birmingham Historical Journal* 2, no. 1: 1–29.

Scamuzzi, E. 1947. Fossile Eocenico con Iscrizione Geroglifica rinvenuto in Eliopoli. *Bolletino della Societa Piemontese di Archeologia e di Belle Arte*, n.s., 1:11–14.

Scilla, A. 1747. *De corporibus marinus lapidescentibus quae defossa reperiuntur*. Rome: Antonii de Rubeis.

Schlosser, W. 2005. Die Himmelsscheibe von Nebra—Sonne, Mond und Sterne. *Acta Historica Astronomiae* 25:27–65.

Schmandt-Besserat, D. 1998. A stone metaphor of creation. *Near Eastern Archaeology* 61:109–17.

Schmidt, B. 1970. Die späte Völkerwanderungszeit in Mitteldeutschland. *Veröffentlichungen des Landesmuseums für Vorgeschichte in Halle* 25:1–102.

Schouten, J. 1968. *The Pentagram as a Medical Symbol*. Nieuwkoop: De Graaf.

Schumacher, I. M. 1988. *Der Gott Sopdu—Der Herr der Fremlander*. Fribourg: Academic Press.

Sedman, L. 2002. The small finds. In *Busayra Excavations by Crystal-M. Bennett*, ed. P. Bienkowski, 353–429. British Academy Monographs in Archaeology 13. Oxford: Oxford University Press.

Sehested, F. 1878. *Fortidsminder og Oldsager fra Egnen om Broholm*. Copenhagen: C. A. Reitzel.

Shaw, I., and P. Nicholson. 1995. *British Museum Dictionary of Ancient Egypt*. London: British Museum.

Sieveking, G., and M. B. Hart, eds. 1986. *The Scientific Study of Flint and Chert: Fourth International Flint Symposium, Brighton Polytechnic, 1983*. Cambridge: Cambridge University Press.

Smith, W. G. 1894. *Man, the Primeval Savage*. London: Edward Stanford.

Soles, J. S., A. M. Nicgorski, T. Carter, and M. T. Soles. 2004. Stone objects. In *Mochlos IC. Period III, Neopalatial settlement on the coast: The Artisan's Quarter and the Farmhouse at Chalinomouri; The Small Finds*. Prehistory Monographs 9. Philadelphia: INSTAP Academic Press.

Sørensen, V. 1994. A er ett ræj i Torrenvejr—for a haar en Torrenstien i æ Lomm!—lidt om echinittens brug og navne. *Ord and Sag* 14:43–53.

Spence, K. 2000. Ancient Egyptian chronology and the astronomical orientation of the pyramids. *Nature* 408:320–24.

Squire, C. 1905. *Mythology of the British Isles: An Introduction to Celtic Myth, Legend, Poetry and Romance*. London: Blackie and Son.

Stanek, K. 1999. Wisiory opasane odmiany wschodniej w środkowoeuropejskim Barbaricum. In *Comhlan: Studia z archeologii okresu przedrzymskiego i rzymskiego w Europie rodkowej dedykowane Teresie Dąbrowskeij w 65. Rocznicę urodzin*, 331–36. Warsaw: Fundacja Przyjaciół Instytutu Archeologii Uniwersytetu Warszawskiego.

Stevenson, R. B. K. 1967. A Roman-period cache of charms in Aberdeenshire. *Antiquity* 41:143–45.

Stone, B. 1974. *Sir Gawain and the Green Knight*. 2nd ed. London: Penguin.
Struve, K. W. 1955. *Die Einzelgrabkultur in Schleswig-Holstein und ihre kontinentalen Beziehungen*. Neumünster: K. Wachholtz.
Suetonius (ca. 69–ca. 122). *The Twelve Caesars*. Translated by Robert Graves. London: Cassell.
Swanton, M. 1975. *Anglo-Saxon Prose*. London: Dent.
Szumowski, G. 1955. Les perles anciennes en oursins fossils. *Notes Africaines: Bulletin d'information et de correspondance de l'Institut Français d'Afrique Noire* 66:33–34.
Taylor, P. B., and W. H. Auden, eds. 1969. *The Elder Edda: A Selection*. London: Faber and Faber.
Thenius, E., and N. Vávra. 1996. *Fossilien im Volksglauben und im Alltag: Bedeutung und Verwendung vorzeitlicher Tier- und Pflanzenreste von der Steinzeit bis heute*. Frankfurt am Main: Waldemar Kramer.
Toms, H. S. 1926. Shepherds' Crowns. *Downland Post*, September 1, 1926, 264–65.
———. 1940. Shepherds' Crowns in Archaeology and Folklore. Unpublished presentation to the meeting of the Brighton Natural History Society, January 6.
Tyack, G. S. 1899. *Lore and Legend of the English Church*. London: William Andrews.
Tylor, E. B. 1922. Augustus Henry Lane Fox Pitt-Rivers. *Dictionary of National Biography* 22 Suppl., p. 1141.
Van Buren, E. D. 1945. *Symbols of the Gods in Mesopotamian Art*. Rome: Pontificium Institutum Biblicum.
Vértes, I. 1964. *Tata: Eine Mittelpaläolithische Travertinsiedlung in Ungarn*. Budapest: Akadémiai Kiadó.
Vesly, L. de. 1903. Fouilles dans la forêt de Rouvray, en 1903. *Bulletin de la Commission des Antiquities* 13:84–88.
Vogel, C. J. de. 1966. *Pythagoras and Early Pythagoreanism: An Interpretation of Neglected Evidence on the Philosopher Pythagoras*. Assen, The Netherlands: Van Gorcum.
Wagner, W., and M. W. Macdowall. 1891. *Asgard and the Gods: The Tales and Traditions of Our Northern Ancestors Forming a Complete Manual of Norse Mythology*. London: Swan Sonnenschein.
Waller, R. 1705. *The Posthumous Works of Robert Hooke, M.D., S.R.S., Geom. Prof. Gresh. Ec. Containing his Cutlerian Lectures, and other Discourses read at the meetings of the illustrious Royal Society*. London. Facsimile reprint, New York: Arno, 1969.
Webber, F. R. 1992. *Church Symbolism: An Explanation of the More Important Symbols of the Old and New Testament, the Primitive, the Mediaeval and the Modern Church*. 2nd rev. ed. Detroit: Ornigraphics.
Weiner, J. S., K. P. Oakley, and W. E. Le Gros Clark. 1953. The solution of the Piltdown problem. *Bulletin of the British Museum of Natural History, Geology* 2:139–46.
Wendt, H. 1968. *Before the Deluge*. London: Victor Gollancz.
West, S. 1988. *The Anglo-Saxon Cemetery at Westgarth Gardens, Bury St Edmunds, Suffolk*. East Anglian Archaeology, Report no. 38. Bury St Edmunds, UK: Suffolk County Planning Department.
White, M. J. 1997. The earlier Palaeolithic occupation of the Chilterns (southern England): Re-assessing the sites of Worthington G. Smith. *Antiquity* 71:912–31.
White, R. 1992. The earliest images: Ice Age "art" in Europe. *Expedition* 34:37–51.
———. 1993. Technological and social dimensions of "Aurignacian-age" body ornaments across Europe. In *Before Lascaux: The Complex Record of the Early Upper*

Paleolithic, ed. H. Knecht, A. Pike-Tay, and R. White, 277–99. Boca Raton, FL: CRC Press.

White, S. 2001. John Henry Pull: A biography. In *Rough Quarries, Rocks and Hills: John Pull and the Neolithic Flint Mines of Sussex*, ed. M. Russell, 5–12. School of Conservation Sciences Occasional Paper 6, Bournemouth University. Oxford: Oxbow Books.

Whittle, A. 1996. *Europe in the Neolithic: The Creation of New Worlds*. Cambridge: Cambridge University Press.

Whittle, A. W. R., J. Pollard, and C. Grigson. 1999. *The Harmony of Symbols: The Windmill Hill Causewayed Enclosure, Wiltshire*. Oxford: Oxbow Books.

Williams, H. 2008. Anglo-Saxonism and Victorian archaeology: William Wylie's *Fairford Graves*. *Early Medieval Europe* 16:49–88.

Wilson, D. R. 1975. "Causewayed camps" and "interrupted ditch systems." *Antiquity* 49:178–86.

Wirtzung, C. 1617. *The General Practise of Physicke. Conteyning all inward and outward parts of the body, with all the accidents and infirmaties that are incident upon them, even from the crowne of the head to the sole of the foote. Also by what meanes (with the help of God) they may be remedied: very meete and profitable; not only for all Physitions, Chirurgians, Apothecaries, and Midwives, but for all other estates whatsoever; the like whereof as yet in English hath not been published*. Translated by Iacob Mosan. London: Thomas Adams.

Woodward, J. 1725. *A Catalogue of the Foreign Extraneous Fossils*. Unpublished catalogue, Sedgwick Museum, Cambridge, dated July 10, 1725.

———. 1728. *An Attempt Towards a Natural History of the Fossils of England: In a Catalogue of the English fossils in the collection of J. Woodward*. Vol. 2, F. Fayram, at the Royal Exchange, J. Senex in Fleet St and J. Osborn and T. Longman in Paternoster-Row, London.

———. 1729. *An Attempt Towards a Natural History of the Fossils of England: In a Catalogue of the English fossils in the collection of J. Woodward*. Vol. 1, F. Fayram, at the Royal Exchange, J. Senex in Fleet St and J. Osborn and T. Longman in Paternoster-Row, London.

Worschech, U. F. Ch., Y. Rosenthal, and F. Zayadine. 1986. The Fourth Survey in the North-west Ard el-Kerak and Soundings at Balù. *Annual of the Department of Antiquities of Jordan* 30:285–310.

Wylie, W. 1852. *Fairford Graves*. Oxford: Parker.

Wynn, T. 2002. Archaeology and cognitive evolution. *Behavioral and Brain Sciences* 25:389–402.

Wyse, P. N., and M. Connolly. 2002. Fossils as Neolithic funereal adornments in County Kerry, south-west Ireland. *Geology Today* 18:139–43.

Zammit, T. 1930. *Prehistoric Malta: The Tarxien Temples*. Oxford: Oxford University Press.

Zammit Maempel, G. 1982. The folklore of Maltese fossils. *Papers in Mediterranean Social Studies* 1:1–29.

INDEX

Abbott, George, 136
abstract thought, 7–9, 25–26, 138
Acheulian tools, 8, 22–23, 28, 30–33, 35, 88, 99
aesthetics, 9
Africa, 5, 12, 21, 25, 30, 82, 153, 200
Agaricus, 97
Agassiz, Louis, 207
Agricola, Georg, 191
'Ain Ghazal, 157–65
Ajlun Group, 158
Akerman, John Yonge, 210
Alberti, Leon Battista, 49
Albertus Magnus, 48
Algeria, 163
Alexandria, 169
Alfred the Great, King, 205
Allen, Don, 206
Altamira, 36, 224
Amenhemet III, 173, 180
Amenhemet IV, 173
Amenhirkhopshef, 171
Amenhotep, 174
Amenophis II, 176
Amman, 155–56, 158, 165
ammonites, 29, 81–82, 111, 164
amulets, 26, 82, 119, 145–46, 151, 166, 194–96, 199, 209, 215
Anglo-Saxons, 133, 146, 150, 200, 202, 204; barrows, 129; burial practices,197, 198–99, 200; poems, 128, 139, 203–4; sea urchin, 210; star, 248n18
Angmering, 124
Angusciola, Spirito, 50
Ankhnesneferibe, 162
An-Ki, 162
Anubis, 174
apotropaism, 13, 130, 216, 218, 221, 224
Arbor Low, 219
Arcy-sur Cure, 26
Arietites, 81
art: Bronze Age, 196; Neolithic, 26–27; Paleolithic, 24–27, 31, 36, 224
Asgard, 149
Ashey Down, 138–40
Ashmansworth, 221, 231
Asterias, 53
Aswan, 171
Asyu, 171
Atropos, 202, 206, 247n5
Augustus Caesar, 169
Aurvandel, 150
Australia, 9, 215
Avicenna, 48
Azekah, 215

Baghdad, 217
Balanocidaris, 38
Baldur, 151–52
Ballycarty, 81–82

Ballymena, 119
Barrow, Benjamin, 139
barrows: Anglo-Saxon, 129; Beowulf's, 139, 203–4; Bronze Age, 65–66, 86, 94, 100–102, 104–5, 124, 129, 138–40, 144; habitation of the *sidhe*, 129; myths, 202; Neolithic, 76, 81–82, 86, 88, 129; nineteenth-century excavations, 114–15; quoit support, 241n1; remembrance, 86, 129
Bayerisches Nationalmuseum, 218
Beachy Head, 122
Beaker culture, 76–77, 240n29
beggarman's knee-caps, 127
belemnites, 37–38, 91, 119, 142, 238n18, 243n12
Belgium, 69, 87, 89–90
Bellucci, Giuseppe, 89
Beltout, 122
benben stone, 183–84
Bennett, Crystal-M., 163–65
Bentheim, 218
Benu, 183, 184
Beowulf, 128, 139–40, 201, 203–4, 243n34
Bible, 164, 217
Bienkowski, Piotr, 164
Biorh, 129
Birka, 195, 196
bishop's miters, 127
bishop's knees, 217
bivalves, 31, 46, 81, 91, 111, 166
blackleading, 124, 242n14
bladder ailments, 38–39
Blinkenberg, Christian, 109, 141–45, 152
Blunt, William, 106
Boccone, Paolo, 50
Boethius, 205
Bolbaster, 35
Bonaparte, Napoleon, 178
Bornholm, 142, 196
bower bird, 9
Bozrah, 164
brachiopods, 35, 37, 46, 81, 91, 111
Brahe, Tycho, 216
Bralund, 203
Bramdean, 222–23
Brighton, 57, 66–68, 70–71, 100, 122, 124, 127, 130, 144, 242n16

Brighton Museum, 64, 68, 73, 77, 121–22, 126, 131
Brittany, 86, 122, 225
Brontia, 191–92
Bronze Age: Beaker culture, 240n29; Crete, 165; Cyprus, 209; Denmark, 144, 147; England, 92, 103, 105–6, 122, 129, 138, 191, 225; France, 86, 124, 225; Israel, 168; Malta, 167
brooches: bronze, 199; saucer, 210; stylized urchin, 195–96
Brookes, Richard, 192
Brown, Gary, 30
Bruele, Gaultherus, 38
Buckland, William, 47
burial chambers, 13–14, 75, 82, 84, 176, 177
burials: Anglo-Saxon, 199; Bronze Age, 92, 94, 102, 105, 124, 129, 138–39, 194, 225; Christian, 198; cremation, 93, 136; Iron Age, 138, 144, 163, 200; Neolithic, 14, 66, 70, 72, 74–77, 82, 84, 86, 129, 194, 217, 225; practices, 5, 7, 13, 65, 114, 115, 138, 175–77; Romano-British, 114, 118; Viking, 195–96
Busayra, 163–165
button-stones, 55–56, 59, 85

Caburn hill fort, 69
Cairo, 169
Caligula, 186
Cam River, 218
Cambridge, 53, 69, 141, 211
Camp-à-Cayaux, 89
Canaanite, 168
cannibalism, 44, 75
cap-stone, 127
Capernaum, 215
Casaubon, Meric, 53
cathedral: Autun, 126; Cologne, 97; drains, 96; Durham, 114; five-pointed stars, 215; medieval, 93; Melbourne, 215; Worcester, 97
Catherine de Medici, 49
causewayed camps, 68–70, 76, 122
Celts, 103, 105, 128–29, 131, 137, 145–46, 191–194, 200–1
Chalcolithic, 168
chalk, 1, 5, 17, 19, 20, 27, 42, 45–47, 52, 54–57, 59–61, 63, 67, 69, 72–75, 80, 83,

87, 92, 94, 101–2, 109–11, 114–15, 118, 121, 126–27, 138–40, 157–59, 191–92, 197, 199, 202, 244n30
Champollion, Jean-François, 178, 183
Charente, 34–35, 37, 208
Charente-Maritime, 85–86, 208
Charles II, 39
Chauvet, 26, 36
chert, 32–33, 58, 81
Chilterns, 94, 99, 100
Christian: beliefs, 11, 41, 151–52, 174, 197, 199, 203–4, 206, 212, 215–17, 219–20; burials, 198
church: Bramdean, 222–24; five-pointed stars, 215, 217–18, 220; flint, 59; medieval, 14, 18, 220–21; Saxon, 220–21; St. Peter's, Linkenholt, 17–19, 24, 218, 220–22, 227–31
cidaroids, 38, 48, 50, 189–90, 194, 209
Clark, Rundle, 182, 184
Claudius, 185–88, 194
clay tablets: Assyrian, 217; Proto-Elamite, 217
Clotho, 202, 206
Clypeus, 110, 241n1
Coatmocum-en-Brennelis, 86
Coenholectypus, 158–59, 161, 164–65, 168, 215
cognitive evolution, 8, 25, 27–28, 30, 34, 40
colepexies' heads, 127
collecting: fossils, 5, 14, 40, 46–47, 57, 64, 81, 90, 199; fossil urchins, 5, 12, 34–35, 41, 55, 57, 61, 78, 130, 153, 168, 174, 216; fungi, 97; habits, 5–6, 9–13, 34, 40, 43–45, 57, 59, 64, 74, 77, 80–90, 98, 110–11, 117, 171
Constantine I, 215
corals, 29, 32, 119
Cotswolds, 111
Cranborne Chase, 114, 116, 122, 130
cremation, 11, 77, 93, 136, 138, 140, 195, 200
Cretaceous, 60, 158, 164, 191, 208
Crete, 165, 202, 224
crinoids, 35, 53, 81
Cro-Magnons, 36
Cuckfield, 65
cultural evolution, 88
Curwen, Cecil, 14, 68–77, 159
Cyclaster, 34

cymbelstern, 216
Cyphosoma, 200
Cyprus, 165, 209
Czech Republic, 37

Darwin, Charles, 43, 111–13
Daviau, Michèle, 168
Davidson, H. R. Ellis, 201
Dead Sea, 158, 164
Decima, 202
Deir el Medina, 171
Demnard, François, 208
Denmark, 76, 84, 138, 140–41, 143–45, 149–50, 195–96
Derbyshire, 219
destiny, 9, 28, 193, 200–203, 205, 207–10, 220, 224–25
Devil's Doors, 220–21, 230
Diodorus Siculus, 193
Dioscorides, Pedanius, 38
Dolní Vestonice, 37
domestication, 79, 83, 156–57, 205
Dom River, 37
Donar, 145–46
Dorset, 11, 33, 114, 116, 119, 125, 127, 130–32, 142, 145
Douglas, James, 198–99
Downes, Dorothy, 24
dragons, 203–4
Drayton, Michael, 206
druids, 187–88, 190, 192–94, 203
Duat, 174, 178–81
Duderstadt, 83
Duffin, Chris, 39, 126
Dumuzi, 162
Dunstable, 91–94, 96, 99, 100, 102–4, 106–7, 118
Dyer, James, 102

Earendel, 150
Echinocorys, 20, 56, 57, 72, 84–85, 87, 89, 101, 109, 121, 125–32, 147, 151, 158, 188, 197, 199, 224, 230, 241n27
Echinolampas, 172, 174
Edom, 164
Edomites, 163–64
Egypt, 12–14, 47, 162, 164, 168–84, 202, 217, 225, 245n19, 245n31, 249n18
Eiseley, Loren, vi, 31

El Hammamia, 171
Elizabeth I, 39
Elliot, Jannion Stele, 100
Endimion and Phoebe, 206
England, 5, 11, 14, 27, 31, 43, 45, 47, 52–53, 59, 64, 66, 68–69, 76–77, 79–81, 83–84, 87, 91, 105, 110, 112, 114, 118, 120–21, 125, 127–28, 133, 138, 141, 145, 149–50, 158, 169, 174, 183, 199–200, 202, 206, 210, 218–19, 238n34
engravings, 7, 25–26, 102, 148, 215
Ennead, 170, 175
Entoloma, 98
Eoanthropus, 88
Er, Myth of, 205
Euzenberg, 83–84
Evans, George, 128
Evans, Sir John, 99

Fairford, 210
fairies, 127–29, 243n12
fairy loaves, 12, 119–21, 123–25, 127–29, 130–31, 133
fairy's night-caps, 127
fairy weights, 127
Falster, 142, 144
Fate, 9, 28, 178, 200, 201–7, 210, 220
Faust, 211, 217
Fenland, 218
Field of Offerings, 175
Field of Reeds, 175
Field of Rushes, 180
Feliks, John, 26
fertility god, 146
fertility symbols, 14, 76, 87, 146, 158, 161–62, 164–65
Five Knolls Tumuli, 94
flint formation, 46, 59–61
flint implement: Acheulian, 20–24; Bronze Age, 93; Mousterian, 35–36; Neolithic, 80–81, 84, 87, 89–91, 136–37; Paleolithic, 20–24, 28, 30–31, 33–35, 41, 98–99
flint mines; Blackpatch, 76; Cissbury, 76; ; Grimes Graves, 87; Harrow Hill, 69; Jablinnes, 87; Kvarnby, 87; Serbonnes, 87; Spiennes, 87
flint urchins, 5, 8, 14, 18–19, 33, 43, 48, 54, 56–57, 63, 83–84, 87, 89–91, 119, 123–25, 127, 194, 197–98, 200, 210, 221, 224, 227, 229
flint uses, 58–59, 67, 80, 84, 87, 91
folklore, 11–14, 77, 87–89, 119–22, 128, 132–33, 141, 145, 147, 189, 242n16, 243n12
Fôret de Montgeron, 89
Forêt de Rouvray, 85, 140
Fossey, Frederick, 95, 101, 104
Fracastoro, Girolamo, 49
France, 24, 26, 33–37, 79, 85–87, 89, 124, 126, 140, 183, 187–88, 190, 194, 197, 200, 208, 224
Freyja, 148
Frigg, 146, 148
Frilford, 200
Funen, 142, 144, 195
fungi, 23, 93, 95–98

Galerites, 85, 191, 195, 197, 210
Gaptooth, 149, 196
gastropods, 37, 81, 166
Gaza, 168
Geb, 162, 170
George IV, 64
Germany, 53, 69, 83, 85, 105, 148, 191, 195, 200, 218
Gesner, Conrad, 48, 52–53, 188–89, 191–92
Gezer, 215
Glastonbury Tor, 81, 82
Glossoptera, 48
Gloucestershire, 145, 150
Goodyer, John, 38
grave goods, 11, 66, 72, 74, 81, 105, 118, 136, 138–39, 195, 198, 200
Graves, Robert, 202
Gravettian, 37
Greece, 141, 202, 216
Greeks, 47, 169, 175, 202, 217
Greenwell, William, 114–15
Grendel, 204
Grotte du Trilobite, 26
Gryphaea, 81

Hackney Downs, 99
Haithabu, 196–97, 217
Hampshire, 18, 20, 33, 125, 127, 145, 198, 221–23
Hanover, 218

Harrow Hill, 69
haruspices, 194
Hathor, 168, 173
Haute-Saône, 124
Heaney, Seamus, 203
heart urchin, 56, 111, 127, 158, 164–65, 168, 188, 190, 200, 247n5
Hebrews, 217
Helgi, 202–3
Heliopolis, 169–71, 173–76, 181–83, 224
Heliopolitan theology, 162
helmet-stones, 55–56, 59
Hermopolis Magna, 171
Herodotus, 47, 182–83
Heterodiadema, 164, 168
hieroglyphs, 12, 171–73, 178
Highbury New Park, 99
Hittites, 146
Höd, 151
holectypoids, 159
Holland, Henry, 65
Hollingbury Castle hill fort, 122
Homo erectus, 8, 21–23, 28
Homo habilis, 7
Homo heidelbergensis, 7, 8, 21–23, 28, 33, 35–37, 41
Homo neanderthalensis, 7, 34, 35–37
Homo sapiens, 9, 10, 21, 23, 36, 88
Honnecourt, Villard de, 215
Hooke, Robert, 54–57, 59, 85
Horus, 170, 175, 179–81, 184
Hove, 66, 68, 131
Huguenots, 49
Hungary, 26
Huon, Christina, 122
Hypogeum, 166

Ice Age, 21, 36, 67, 197
Iguanodon, 65
Imperato, Ferrante, 50
Inanna/Ishtar, 162
Indra, 146
Iraq, 217
Ireland, 26, 81, 112–13, 119, 138
Iron Age: Crete, 166; Denmark, 144, 195; England, 41, 69, 115, 122, 137, 144, 184; Europe, 200; France, 85–86, 208, Jordan, 160, 163–65, 208; Near East, 163; Switzerland, 137–38

Isastraea, 32
Isis, 170, 175, 179–81
Isle of Wight, 54, 56, 59, 127, 138–40
Israel, 39, 157, 167, 215
Israelites, 163–64
Italy, 49, 89, 119, 141, 171–72, 219
Ivanov, Sergei, 214

Jemdet Nasr, 217
Jerusalem, 215
jew stones, 38–39
Jordan, 39, 155, 158, 160–61, 163–65, 167–68, 208, 210, 215
Jörth, 146, 148
Juick, 86
Juliet, 206
Jutland, 138, 142–44, 197

Kaw el Kebir, 171
Kent, 22–23, 52, 128, 135, 137, 198, 200
Kerak, 155, 160
Keraunos, 141
Kha, 171
Khaemwaset, 171
Ki, 162
King Lear, 206
Knauthain, 218
Kommos, 165–66
Kostenski, 37
Kremlin, 214
Ksar 'Akil, 38

La Gravette, 37
La Motte St. Jean, 84–85, 140
La Placard, 37
La Roche-au-Loup, 34
La Tène culture, 137–38
Lachesis, 202, 206
Landes, 34
Langeland, 142
Lapides sui generis, 53
lapis judaicus, 38–40
Lascaux, 26, 36–37, 224
Lay of Harbard, 150
Lebanon, 38
Leipzig, 218
Leonardo da Vinci, 13, 49, 161, 216, 219
Lévi-Strauss, Claude, 32
Libya, 163

Limassol, 209
limestone: Carboniferous, 81, 219; Cretaceous, 158, 164; Egyptian, 179, 182; Eocene, 47, 171; figurines, 157–58, 160; Miocene, 50; Jurassic, 111; Sinai, 174; use in clothing, 37
Lindos, 141
Linear pottery culture, 84
Linkenholt, 17–20, 24, 33, 218, 221–22, 227–31
Linnaeus, Carl, 10
Linthia, 168
literature: Elizabethan, 206; Jacobean, 206
Loki, 151–52, 244n33
Lolland, 142, 144, 195
long barrow, 81
Lovett, Edward, 145, 244n29
Lyell, Charles, 47

Macbeth, 202
Magdalenian, 26
magic, 10, 28, 37–38, 77, 88, 112, 119, 129, 146, 167, 180–81, 187, 189, 192, 194, 198, 200, 205, 209, 212, 216, 243n12
Maimonides, Moses, 39
Malta, 50, 112, 165–67
mammoth hunters, 37
Mantell, Gideon, 46–47, 65, 127–28
Maramius, 98
Marissa, 216
masons, 49, 215–16
Maud, 93, 101–7, 109, 111, 117, 128, 158–59, 174, 179, 184, 200, 224–25
Mayor, Adrienne, 132
Mecaster, 158, 164
medieval, 13–14, 18, 39, 49, 52, 114, 127, 137, 188, 212, 215, 217, 220–22, 230–31
Mediterranean, 12, 14, 50, 153, 156, 160, 163, 165, 202
Melbourne, 215
Mephistopheles, 211, 217
Mercati, Michele, 48, 189, 191
Merenre, 176
Merneptah, 175
Merovingian, 195, 197
Mesolithic: fossil collectors, 29, 40, 41; Sweden, 40
Mesopotamian: deities, 162; myths, 162, 217
Messenius, Johannes, 149

Metzger-Krahé, Frauke, 197
Meuris, Roland, 89–92
Micraster, 33, 37, 56–57, 63, 85, 89–90, 101, 111, 125, 127–28, 136–37, 158, 188, 190, 192, 197–98, 200, 230, 241n27
Middle Ages, 48, 126, 127, 147, 149, 217–18
Milankovitch cycles, 60
Minoan, 165
Mittelberg, 105
Mjöllnir, 66, 146–47, 149, 150–53, 244n33
Moerae, 202
Mont Vaudois, 124
Moon goddesses, 202
Morta, 202
Mousterian, 26, 34–36, 237n6
Muensterberger, Werner, 10, 45
Munich, 218
Museo Egizio, 171–72, 182
Mycenean, 141
mythology: Babylonian, 202; British, 128–29, 150; Cretan, 202; Egyptian, 170, 180, 182–84; evolution, 9; fossils, 12–13, 15, 89, 147, 152; Greek, 205; Icelandic, 149, 202; Mesopotamian, 162; Near Eastern, 162, 202; Norse, 13–14, 133, 146, 148–49, 152; Palestinian, 202

Nammu, 162
Naqada, 168
Nash, John, 65
Neanderthals, 7, 26, 34–37, 237n6
Near East, 5, 12–13, 29, 39, 79, 162–63, 200
Necessity, Goddess of, 202–3, 205–6
Nefertari, 171
Negev, 168
Neolithic, 41, 80, 83, 105, 129, 205, 225; Algeria, 163; art, 82; Belgium, 89–91; Cyprus, 165; Denmark, 84; Egypt, 168; England, 41, 68–70, 73–74, 76–77, 80–81, 88, 115, 122, 137, 184, 219; flint mines, 87; fossil collecting, 29, 41, 81, 84, 87–88, 129, 191, 194, 200; France, 84–86, 89, 140, 208; Germany, 83–85; Ireland, 81; Italy, 89; Jordan, 156–58, 160–63, 165, 208, 215; Libya, 163; Malta, 167; Niger, 163; Pre-Pottery, 157; Sudan, 163
Neoplatonism, 52
Nephthys, 170, 175
Néraudeau, Didier, 34–35, 208

Nettesheim, Agrippa von, 216
Nichols, Thomas, 53
Niger, 163
Nile River, 169, 174, 180
Nona, 202
Norfolk, 31, 128, 130, 220
Nornir, 202, 205
Northumberland, 209
nummulites, 26, 47
Nun, 170
Nut, 162, 170, 175, 181

Oakley, Kenneth, 24, 26–27, 29, 37, 87–88, 125–27, 131, 168, 209
obelisks, 169
Oberjohnsdorf, 85
ochre, 25
Odin, 146–49, 151
Offaster, 57
Olaus Magnus, 148–49
ombria, 191–92
On, 169
orcs, 128
Osiris, 170, 175, 178–80
ovum anguinum 48, 185, 187–94, 197, 199, 207
oysters, 35, 50, 198
Oxfordshire, 53, 110, 119, 191

Pakistan, 39
Paleolithic, 12, 29, 34, 38, 40–41, 87; art, 32, 36–37; artifact, 22–23, 34, 84, 88, 90, 99, 160, 208; Czech Republic, 37; England, 22–23, 99–100; fossil collectors, 13, 24–25, 34, 38, 40, 87, 208; France, 34–36; Lebanon, 38; South Africa, 25; Spain, 36, 224; Ukraine, 37
Palissy, Bernard, 49, 53
Parcae, 202, 205
passage tomb, 81–82
Pegiocidaris, 85
pendants: amuletic, 195; belemnite, 37; bound, 195–96; five-pointed star, 42; fossil coral, 119; lignite, 26; magical, 37; Paleolithic, 40; stone, 81
pentagram, 47, 211–12, 214–15, 218–19; 248n3; on amphora, 216; as apotropaic symbol, 216–18; on bed, 218; on buildings, 215, 217–18; on cradle, 218; on pots, 215; on tombs, 216

pentalpha, 214, 216, 218–19
pentangle, 212–14
Pentateuch, 217
personal adornments, 37, 74
Petrie, Flinders, 158, 176–77
pharisee loaves, 127–28
Phillips, John, 110
Phoenix, 170, 182–84
Phymosoma, 194
Piltdown Man, 88
Pitt, George, 114
Pitt-Rivers, Augustus Lane Fox, 14, 109–20
Pitt Rivers Museum, 119–20
Plagiochasma, 168
Plato, 205–6
Pleiades, 106, 149
Pleistocene, 33
Pliny the Elder, 38, 47–48, 131, 185, 188–92, 246n7
Plot, Robert, 53, 110, 127, 191–92
Poetic Edda, 202
Poland, 195
Poplin, Francois, 34
Porosphaera, 100
poundstones, 110
protozoans, 26
Ptilonorhyncus violaceus, 9
Pull, John, 14, 76–77
pyramid texts, 169, 176–80, 182–83, 245n19
Pythagoras, 47, 216

quintessence theory, 184, 219
quoitstones, 110, 241n1

Ra, 169–70
Ragnarök, 152
Rahtz, Philip, 81–82
Raleigh, Sir Walter, 206
Ramses II, 171
Ramses III, 168, 171
Ramses V, 168
Ramses VI, 175
Ramses VII, 175
Raphiostoma, 165
Re-Atum, 170, 181, 183
rebirth, 152, 162, 180, 182, 184, 190, 224–25
Red Army, 214
Reichenbach, 85

Renaissance, 47, 188, 191, 202; belief in influence of stars, 206; philosophy, 52, 206
Republic, 205
Richard I, 18
Rizokarpaso, 165
Romano-British, 118, 132, 218
Romans, 10, 14, 69, 85, 93, 130–31, 140, 145, 166, 169, 178, 185–86, 191, 193–94, 197, 202, 204, 208–9, 215
Romeo, 206
Rosetta Stone, 178
Rotherly, 118
Rowe, Arthur, 111
Royal Pavilion, 65
Rudwick, Martin, 48
"Ruin, The," 204
Rushmore, 114, 116, 118
Russia, 37, 150, 195, 214

Sah, 170, 179–81, 184, 200
Saint-Amand-sur-Sevres, 86
Saint-Just-des-Marais, 33, 91
Saintonge, 85
Salisbury Plain, 17
Saône-et-Loire, 84–85, 140
Sapinières, 89
Saqqara, 176–77, 179
Saxon: churches, 220–21; period, 41, 207–8
scallops, 29
Scamuzzi, Ernest, 173
scaphopods, 37
Schiaparelli, Ernesto, 170–73
Schmandt-Besserat, Denise, 162
Schouton, Jan, 162
Schumacher, Inke, 173–74
Scilla, Agostino, 50–52, 60
Scolopendrites, 191
Scott, George Gilbert, 97
sea urchin fossils, 19, 29, 40–44, 46, 49, 55–56, 59–66, 111, 207; as amulets, 86, 145, 194–96, 199; Anglo-Saxon, 41, 198–200; apotropaic power, 13, 130, 147, 199, 216, 218, 221; artificial modification, 13, 158–59, 164–65; with axes, 85, 136–37, 140, 144, 152; in barrows, 14, 86, 129, 139–40, 144; Bronze Age, 86, 101–7, 124, 165, 200; in buildings, 11, 131–32, 164, 197, 218, 231; in burial with a child, 5, 15, 118; in churches, 14, 18–19, 218, 220–23, 230; collecting, 9, 12–14, 20, 27, 47, 54–55, 57, 59, 61, 63–64, 80, 85, 87, 153, 156, 158–59, 174, 182; in cremation, 11, 136–40, 195, 200; Danish folklore, 142–44; drilled, 12, 37, 85, 159–63, 165, 168, 200, 208–10; English folklore, 20, 33, 121–33, 145, 191–92; fertility object, 161–62, 164–65; folklore, 13, 49–52, 77, 88, 119–33, 225; in graves, 5, 11, 40–41, 64, 67, 72–78, 83, 101–3, 118, 152, 168, 184, 191, 197–200, 210; with hieroglyphs, 12, 171–74, 183; influence on Fate, 207–8, 210, 220; Iron Age, 144, 164–66, 200; magical object, 167, 194, 198; medicinal value, 38, 39, 52–53; Merovingian, 195, 197; Mesolithic, 40; myths, 15, 147–48, 150; necklace, 156, 160, 168; Neolithic, 72–77, 81–89, 160–63, 165, 168, 200; in Neolithic tools, 89–91; as *ovum anguinum*, 48, 187–92, 194, 197; in Paleolithic tools, 13, 22, 24, 27–28, 30, 32–35, 40, 42, 88; as personal ornaments, 160, 194–96; and rebirth, 152, 182, 184, 191, 198, 224–25; Romano-British, 116–18; spines, 37–39; spiritual significance, 12, 28, 40, 83–85, 129, 144, 165–68, 174, 200, 208, 210, 225; star symbol, 14, 52, 83, 105–6, 110, 148, 151, 160, 164, 168, 174–75, 181, 212, 215, 217, 224; stylized in brooch, 195–96; in temple/shrine, 6, 164, 166–68; use as spindle whorls, 85, 209–9; Viking, 195–97, 200
Seal of Solomon, 213, 217
Sedgwick, Adam, 47
Sedgwick Museum, 47
Senfore, 173
Senwosret I, 169
Seth, 170, 175, 181
Seti I, 175, 177
Shakespeare, William, 63, 135, 202, 206
sheep's hearts, 33, 127
shepherds' crowns, 12, 19–20, 47, 77, 119–23, 133, 143, 145, 188–89, 225, 230, 231, 244n29
shepherds' hats, 127
shepherds' hearts, 127
shepherds' knees, 127
Shoreditch, 98

Shu, 162, 170
Sidhe, 128–29
Silbury Hill, 81
Sinai, 168, 173–74
Sir Gawain and the Green Knight, 212–15
Sirius, 148, 179–81
Sismondia, 37
Skadi, 149–50
Skuld, 202
skulls: brachycephalic, 105; dolichocephalic, 105
sky disc, 105
Smith, Alan, 14, 33
Smith, Reginald, 137
Smith, William, 110–11, 115, 241n1
Smith, Worthington, 14, 91–107, 109–10, 116–17, 132, 159, 179
Somerset, 81
Sönder Omme, 144
Sopdet, 170
Sopdu, 170, 172–74, 179–84
Soped, 170
Sophet, 170
South Africa, 25
South Downs, 44–46, 67, 69, 76–77, 115
Southborough, 135, 137
Soviet Union, 214–15
Spain, 36, 216, 224
Spencer, Herbert, 113
Sphaerodus 119
spindle whorls, 160, 205, 208–9, 217
Spindle of Necessity, 205
spinning, 60, 205, 208–9, 210
spirituality, 8–9, 176, 218
Spondylus, 31
sponges, 29, 46, 59, 100
St. Paul, 50, 52
Stadeleck, 218
stars, 2, 11, 33, 35, 52, 54, 56, 58, 144, 148, 156, 161–62, 164, 211–12, 214, 217, 219–20, 225; astrological, 52, 206–8, 216, 219–20; astronomical, 8, 13, 30, 52–53, 57, 105–6, 138, 147, 149–50, 155, 160, 168, 174, 176–81, 184, 206, 217, 224–25; in Egyptian burial chambers, 175–76, 179, 217; five-pointed, 5, 8–9, 11–15, 19–20, 23–24, 27–28, 30, 33, 35, 40, 42–43, 52, 61, 63, 82, 89, 110, 125, 147, 151, 159, 162, 168–72, 174–77, 192, 199, 209–12, 214–20, 224–26, 229–31, 248n3, 248n18, 249n18; hieroglyph, 177–79; stone, 53, 83, 101, 147, 155, 171, 176, 180, 183–84, 200, 208, 210
Sternstein, 53
Stoke Newington Common, 99
Stolpe, Hjalmar, 195
Stoney Littleton, 81
Strabo, 191
Stradbroke, 220
Studland, 132, 142, 144, 217–18
Sudan, 163
Suetonius, 185–86
Suffolk, 119, 128, 199
sugar loaves, 128, 130
Sumerian deities, 162
St. John's College, Cambridge, 211, 216
Sturluson, Snorri, 151
Sussex, 44, 46, 54, 65–66, 69–70, 74, 77, 87–88, 115–16, 119, 122–25, 127–28, 130–31, 145, 221
Swanscombe, 22–24, 28, 30, 32, 34, 87, 90
Swanscombe Man, 23, 230
symbolism, 12, 25, 28, 75, 88–89, 148, 151, 161, 175, 181, 183, 212, 215–16, 218, 225
symmetry, 7–8, 22–25, 27, 32, 34, 36, 40, 56, 63, 82

Tanarus, 145
Tarxien, 166–67
Tata, 26
Tefnut, 170
Tel Zakariya, 215
Tell el-Ajjul, 168
Tell Jawa, 165
Tercis, 34
Theophrastus, 47
The Trundle hill fort, 69
Thjazi, 149
Thor, 13, 66, 145–53, 196
Thrym, 152
Thunderbarrow, 69
thunderbolt, 119, 145–46, 150, 244n30
thundergod, 149
thunderstones, 2, 3, 47, 133, 137, 141–46, 152–53, 188, 191, 196–97, 200, 225
thunderweapons, 141
Thunor, 154–56
Tilgate Forest, 46

Timaeus, 206
Timna, 168
tin tabernacle, 222–23
Tja-nefer, 172–74, 176, 178–79, 181–84, 224–25
toadstones, 119
Tolkien, J. R. R., 128
Toms, Herbert, 14, 68, 70–71, 77, 120–27, 130–31, 133, 138, 141, 145, 189, 193, 242n7, 242n16
Toothgnasher, 149, 196
Tortosa, 216
trilobites, 26
Tudor, 126, 209
Tumulus de la Fourcherie, 86
Tumulus de Poiron, 86
Turton, William, 10
Tuthmoses III, 169
Tylor, Sir Edward Burnett, 112
typology, 112

Ukraine, 37
Umbria, 89
Unas, 176, 180
Universal Deluge, 38, 54, 198
Uppsala, 148–49
Ursa Major, 148–50
Ursa Minor, 149

Verdandi, 202
Viking age, 146, 151, 195–97
Vikings, 14, 148, 195
Vis plastica, 48
visual symbolism, 25

Vitruvian Man, 13, 40, 161, 216, 219
Vocontian knight, 188, 193–94, 207

Wadi ath-Thamad, 168, 210
Wadi Mujib, 155
Wallace, Alfred, 111
"Wanderer, The," 204
Wandil, 150
weaving, 203, 205, 208–9
Werethekau, 174
West Tofts, 31
Westgarth, 199
Westminster Abbey, 97
White, Mark, 99
Whitehawk, 68–70, 73–77, 80, 83–84, 86, 88, 105, 118, 122, 184
Whittle, Alasdair, 69, 82
Wiccan, 218
Wiltshire, 76, 114, 125, 130, 145
Windmill Hill, 76
Wirtzung, Christopher, 39
Woodhouse, 209
Woodward, John, 38, 52, 238n24
Wulfstan, Archbishop, 206
Wylie, William, 210
Wyrd, 201–5, 247n3

Xanthus, 47
Xenophanes, 47

Young, Thomas, 183

Zammit, Themistocles, 166–67
Zeus, 141, 202